THE COCONUT
Phylogeny, Origins, and Spread

T0282366

THE COCONUT
Phylogeny, Origins, and Spread

N MADHAVAN NAYAR

Amsterdam • Boston • Heidelberg • London
New York • Oxford • Paris • San Diego
San Francisco • Singapore • Sydney • Tokyo

Academic Press is an imprint of Elsevier

Academic Press is an imprint of Elsevier
125 London Wall, London EC2Y 5AS, United Kingdom
525 B Street, Suite 1800, San Diego, CA 92101-4495, United States
50 Hampshire Street, 5th Floor, Cambridge, MA 02139, United States
The Boulevard, Langford Lane, Kidlington, Oxford OX5 1GB, United Kingdom

Notices
Knowledge and best practice in this field are constantly changing. As new research and
experience broaden our understanding, changes in research methods, professional practices,
or medical treatment may become necessary.

Practitioners and researchers must always rely on their own experience and knowledge in
evaluating and using any information, methods, compounds, or experiments described
herein. In using such information or methods they should be mindful of their own safety
and the safety of others, including parties for whom they have a professional responsibility.

To the fullest extent of the law, neither the Publisher nor the authors, contributors, or
editors, assume any liability for any injury and/or damage to persons or property as a
matter of products liability, negligence or otherwise, or from any use or operation of any
methods, products, instructions, or ideas contained in the material herein.

Library of Congress Cataloging-in-Publication Data
A catalog record for this book is available from the Library of Congress

British Library Cataloguing-in-Publication Data
A catalogue record for this book is available from the British Library

ISBN: 978-0-12-809778-6

For information on all Academic Press publications
visit our website at https://www.elsevier.com/

  Working together
to grow libraries in
Book Aid
International developing countries

www.elsevier.com • www.bookaid.org

Publisher: Nikki Levy
Acquisition Editor: Nancy Maragioglio
Editorial Project Manager: Billie Jean Fernandez
Production Project Manager: Susan Li
Designer: Maria Inês Cruz

Typeset by TNQ Books and Journals

CONTENTS

PREFACE

Cocos nucifera, the coconut, is one of the three most economically important palm species in the world, a status shared with African oil palm (*Elaeis guineensis*) and date palm (*Phoenix dactylifera*). The coconut not only features prominently in the export economy of many developing nations in the tropics, but it is an important resource at the subsistence level, not only for food, but for fiber, as well as construction material. Some would call the coconut the true "tree of life." With a seed superbly evolved for oceanic dispersal, today the coconut can be found along the littoral strand of all continents that extend into the tropics.

Despite its critical global role as a sustainable natural resource for the food, cosmetic, health, and manufacturing industries, the natural history of *C. nucifera* remains something of an "abominable mystery," to borrow from Charles Darwin. The precise phylogenetic position of the coconut to other palms remains ambiguous, though it would appear that the genera of closest relationship are South American, classified along with *Cocos* in the tribe Cocoseae, subtribe Attaleinae. Of these, it appears that the genera *Attalea* or *Syagrus* may be the best candidates for true sister genera to the monotypic *Cocos*. Despite these deep relationships with South American palms, no one has ever been able to ascertain true nativity for any population of *C. nucifera* in the neotropics. Moreover, although the fossil record has yielded fragments of ancient plant parts attributed to coconut in deposits found in Colombia, India, and even New Zealand, all dated long before the dawn of humanity, there are many who doubt their veracity. The more recent history of the coconut has been deeply influenced by human activity. The history of the plant's domestication is no less, and actually more, controversial than its phylogenetic origins, and all we can ascertain at this point in time is that there were likely multiple areas of domestication at various times in human history.

Dr. N. M. Nayar, for the first time, brings together all of the diverse literature pertaining to the biogeography, phylogeny, and domestication of the coconut. He is well positioned to synthesize the vast literature on the subject, having worked for many years in tropical crop genetic resources and crop resources.

This is an important achievement and comes at the right time, because the ever–advancing technology for molecular genetic analysis may allow us to once and for all solve the "abominable mystery" of the origins of *C. nucifera* in this century.

Alan W. Meerow, Ph.D.
Research Geneticist and Systematist
United States Department of Agriculture
Agricultural Research Service
1360 Old Cutler Road
Miami, Florida 33158-0000, USA
March 23, 2016

ABBREVIATIONS AND DEFINITIONS

ABBREVIATIONS

AFLP	Amplified fragment-length polymorphism
APCC	Asia and Pacific Coconut Community
APG	Angiosperm Phylogeny Group
BCE	Before the Common Era
BI	Bioversity International
BLAST	Basic local-alignment search tool
BP/bp	Before the Present
Cal BCE	Calibrated BCE
Cal BP	Calibrated BP
CE	Common/Christian era
cpDNA	Chloroplast DNA
FBS	Food Balance Sheet
ISSR	Intersimple sequence repeat
ITS	Internal transcriber spacer
Ka/ka/kya	Thousand(s of) years ago
ML	Maximum-likelihood method
MP	Maximum-parsimony method
M tn, M t	Million tonnes
NJ method	Neighbor-joining method
PCA	Principal component analysis
PCR	Polymerase Chain Reaction
RAPD	Random amplified-polymorphism DNA
RBG	Royal Botanic Garden, Kew, Richmond, UK
rDNA	Recombinant DNA
RFLP	Restriction fragment-length polymorphism
SINE	Short interspersed nuclear element
SNP	Single-nucleotide polymorphism
SSR	Single-sequence repeat
UPGMA	Unweighted-pair group method with arithmetic averages
WCSP	World Checklist of Selected Plants

PREHISTORY/HISTORY ABBREVIATIONS

Kyr	Thousand years
yr bp	Uncalibrated radiocarbon years before present
kyr bp	Uncalibrated radiocarbon kyr Before the Present
cal BP	Calibrated radiocarbon years Before the Present
kyr cal BCE	Calibrated radiocarbon kyr BCE
kyr cal AD	Calibrated radiocarbon kyr CE

BCE   Before the Common Era
AD/CE   Historical years AD/Common Era

DEFINITIONS

Acaulescent	Without a visible stem
Armed	Bearing spines
Bootstrap support	An estimate of confidence in an individual clade within a phylogenetic tree
Caducous	Falling off early
Cespitose	Tufted
Clade	A monophyletic group; a branch on a phylogenetic tree
Costapalmate	With the petiole extending into the blade of a palmate leaf; shaped like the palm of a hand
Dioecious	Having male and female flowers on different plants of the same species
Endocarp	The innermost layer of the pericarp
Endosperm	The albumen of a seed; the nutritive body of the palms
Epicarp/exocarp	The outermost layer of the pericarp
Hypaxanthic	Shoots dying immediately after flowering x Pleonanthic
Jacknife support	An estimate of confidence in an individual clade within a phylogenetic tree
Maximum likelihood	A criterion for choosing among different hypotheses of relationships
Mesocarp	The middle layer of the fruit wall; the fibrous matter in the coconut husk
Monoecious	Having separate male and female plants
Microsatellite	A segment of DNA characterized by a variable number of copies of a sequence of five or fewer bases
Monophyletic group	A taxonomic group that includes all descendants of a common ancestor
Monosulcate	In a pollen grain having one groove or furrow
Parsimony	A criterion for choosing among different hypotheses of relationships
Peduncle	The stalk of an inflorescence
Pericarp	The fruit wall
Pinnate	Having separate leaflets along both sides of a common stalk
Polyphyletic	Having more than one ancestral lineage
Protandrous	Flowers in which the pollen is shed before the stigma becomes receptive
Protogynous	Having stigmas becoming receptive before the stamens of the same flower shed their pollen
Resolution	An arrangement of taxa in a phylogenetic tree indicating the ambiguity in a phylogenetic hypothesis
Sister group	The closest relative in a phylogenetic tree
Spicate	spike-like, An unbranched inflorescence
Topology	The arrangement of branches on a phylogenetic tree

GEOLOGICAL TIMESCALE

Classical

ERA
   Period
     Epoch
CENOZOIC (66.0 Mya to present)
   Quaternary (1.6 Mya to present)
     Holocene (10,000 Ya to present)
     Pleistocene (1.6 Mya to 10,000 Ya)
   Tertiary (66.0 to 1.6 Mya)
     Pliocene (5.0 to 1.6 Mya)
     Miocene (23.0 to 5.0 Mya)
     Oligocene (36.0 to 23.0 Mya)
     Eocene (57.0 to 36.0 Mya)
     Paleocene (66.0 to 57.0 Mya)

MESOZOIC (245 to 66 Mya)
   Cretaceous (144 to 66 Mya)
   Jurassic (208 to 144 Mya)
   Triassic (245 to 208 Mya)
PALEOZOIC (600 to 245 Mya)
   Permian (286 to 245 Mya)
   Carboniferous (360 to 286 Mya)
   Pennsylvanian (320 to 286 Mya)
   Mississipian (360 to 320 Mya)
   Devonian (408 to 360 Mya)
   Silurian (438 to 408 Mya)
   Ordovician (505 to 438 Mya)
   Cambrian (600 to 505 Mya)
PRECAMBRIAN (4600 to 600 Mya)
   Proterozoic (2500 to 600 Mya)
   Archean (4600 to 2500 Mya)

Current

ERA
   Period
     Epoch
CENOZOIC (65.5 Mya to present)
   Quaternary (2.59 Mya to present)
     Holocene (11,700 Ya to present)
     Pleistocene (2.59 Mya to 11,700 Ya)
   Neogene (23.03 to 2.59 Mya)
     Pliocene (5.33 to 2.59 Mya)
     Miocene (23.03 to 5.3 Mya)
   Paleogene (65.5 to 23.03 Mya)
     Oligocene (33.9 to 23.03 Mya)
     Eocene (55.8 to 33.9 Mya)
     Paleocene (65.5 to 55.8 Mya)

MESOZOIC (251.0 to 65.5 Mya)
   Cretaceous (145.5 to 65.5 Mya)
     Upper (99.6 to 65.5 Mya)
     Lower (145.5 to 99.6 Mya)
   Jurassic (199.6 to 145.5 Mya)
     Upper (161.2 to 145.5 Mya)
     Middle (175.6 to 161.2 Mya)
     Lower (199.6 to 175.6 Mya)
   Triassic (251.0 to 199 Mya)

TIMELINE: COCONUT PHYLOGENY *VIS-À-VIS* PLATE TECTONICS/ CONTINENTAL DRIFT

Mya	Geological Era	Activity
1100–750 750 300	 Late Paleozoic– Early Mesozoic	• Rodinia breaks into Laurasia and Gondwana • Pacific Ocean born after Rodinia breakup; then known as Panthalassic Sea • Pangaea forms by joining of continental masses • Further breakup in three phases
208–146		• Origin of angiosperms (Niklas, 1992)
200 180 144–118	 Lower Jurassic Lower Cretaceous	• First break up: Pangaea breaks into Gondwana and Laurasia. Gondwana begins to drift southward • Gondwana breaks up into East and West Gondwana • Second breakup: South Atlantic Sea opens: South America separates from Australia + Antarctica • Pangaea begins to break up into Africa, India, and Australia • Results in opening of south Arabian Ocean
141–106 130–85 120	Lower Cretaceous Lower Cretaceous	• Earliest fossil palm material: Costapalmate leaves of *Subaltes caroliensis* (Berry, 1914) • New Zealand + New Caledonia and rest of Zealandica begin to separate from Antarctica, and Australasia, and begin forward moving eastward toward the Pacific • Madagascar + India separate from Gondwana. Madagascar stops movement after locking onto African plate • Third breakup: Laurasia begins to separate from Eurasia; Europe and North America separate resulting in formation of Indian Ocean
120–110		• Earliest monocot fossil (Friis et al., 2004)
117.9–100.0		• Divergence of Arecaceae (Baker and Couvreur, 2013ab)

Mya	Geological Era	Activity
94–66	Upper Cretaceous	• India continues to move northward at 6 cm/year • Australia splits from Antartica and moves northward at 5–6 cm/year
78.3–73.6		• Divergence of subfamily Arecoideae (Baker and Couvreur, 2013a
66.3–63.8		• Divergence of subtribe Attaleinae (Meerow et al., 2014)
60–55		• Atlantic and Indian Oceans continue to expand closing the Tethys Sea
59.4–55.8		• Divergence of tribe Cocoseae (Baker & Couvreur, 2013a, 2013b)
55.8–36.2		• Divergence of subtribe Attaleinae (Baker and Couvreur, 2013a,b)
55	End Paleocene	• Laurasia divides into Laurasia (South America) and Eurasia (−India)
53.3–61.1		• Divergence of Cocoseae (Gunn, 2004)
45		• Australia + New Zealand's connection with Gondwana ends • Indian plate collides with Asia forming the Himalayas • Antarctica gets cooler, Australia gets warmer
43.7		• Divergence of subtribe Attaleinae (Meerow et al., 2009)
42.9–51.0		• Divergence of nonspiny clade (subtribe Attaleinae) (Gunn, 2004)
30	Oligocene	• South America separates from west Antartica; cooling takes place • New Guinea collides on the Australian plate pushing up New Guinea Highlands • Indian plate continues to push Eurasia steadily increasing the height of the Himalayas
23.9–44.4		• Divergence of *Cocos* (Meerow et al., 2009, 2014)
23.6–3.92	Miocene–Plocene	• Divergence of *Cocos* (Baker and Couvreur, 2013a,b)
22.2–26.7	Oligocene–Miocene	• Divergence of *Cocos* (Gunn, 2004)
15		South America connects with North America

Various, mainly Wikipedia and as given in text above.

INTRODUCTION

The coconut palm, *Cocos nucifera* L., Arecaceae, is the most ubiquitous plant in much of the lowland tropics vegetation of the Old World. It is also dominant in more than 30,000 islands that dot the Pacific and Indian Oceans. The coconut has been an integral part of the legends, lores, and lives of the peoples of this vast region for over 100 centuries during the closing millennia of the Holocene Epoch.

In this region, the coconut has been the most useful tree to humans. Every part of the palm used to be put to some active economical use, possibly from the time the modern humans came across this palm. This has been so at least until about the late 1960s—till the beginning of the Anthropocene epoch—when the habits and lifestyles of modern human society began to undergo a major transformation. Even though both the qualitative and quantitative dependence on the coconut have diminished substantially since then, the coconut continues to be the staff of life even now in many of the island communities.

In the industrially developed countries of Europe and North America, the coconut had a golden phase for over 100 years from the mid-19th century. The end of the Second World War and its aftermath had signaled the fall of this crop plant. The coconut palm has continued as an orphan crop ever since.

This is reflected in coconut research and development publications as well. The last two major monographic accounts of the coconut have been Menon and Pandalai (1958). The Coconut Palm. A Monograph and Child (1974). Coconuts. i.e., 40–60 years ago. We may not overlook Fremond et al. (1966). Cocotier, and the popular Food and Agriculture Organization (FAO) handbook, Ohler (1984), The Tree of Life.

In biological sciences, the advent of molecular biology has brought about a sea change since about the last five decades in the approaches, methodologies, and reach that were not hitherto possible with the classical techniques. Nevertheless, true to the orphan status of the coconut, hardly any specifically directed studies of the coconut have been published in the areas involving molecular biology applications.

In the present volume, I have attempted to piece together all the information available on the coconut in various diverse areas as geology, palynology, history, archeology, molecular biology, and taxonomy. I have attempted to collate and synthesize this information in the first seven chapters of the volume—The place of the coconut in the world; early history and lore; taxonomy and intraspecific classification of the cultivated coconut; paleobotany and archaeobotany; phylogeny; and biogeography. Using the information obtained in the areas that are relevant to crop-plant origins, I have attempted in Chapters 8 and 9 a synthesis on the origin of the coconut. It is my expectation that it has been possible to piece together the puzzle relating to the origins of the coconut. At the minimum, the subject will now move beyond the realm of speculation.

Now move beyond the realm of speculation.

In the penultimate Chapter 9, Spread, we can see the continuing uncertainty about even the initial introduction of the coconut into the New World. The complementary poser about the Pre-Columbian presence of the coconut on the Pacific coast of America continues as a smoldering problem. The volume ends with Chapter 10, Afterward, with the poser, Has the coconut a future? I am moderately optimistic that this Tree of Life, this Staff of Life, and this Keystone Species—of the several thousands of islanders of the Indian and Pacific Oceans and also some communities in south and southeast Asia—for whom the coconut is also a secondary staple—can gave a good future with some external support. This is because the coconut palm is the most useful and versatile multipurpose plant of the present times. I cannot think of the coconut in any other way, as I was born and spent almost half of my life in the state of Kerala, India. Kerala means the land of the coconut!

Assembling the literature needed for this venture has been a truly uphill task, located as I have been in an isolated city in a third world country. Most of those living and working in North America, Europe, Japan, Australia, and New Zealand will never be able to comprehend the tribulations, time, and expenses involved in assembling the needed literature. Although I had access to the journals and other publications available online, the difficulties lie in procuring the rest. Some of these old publications are available from bookstores like Amazon, Abe, etc. Although cost of the publications may look very modest, the intimidating factor is the packing and forwarding charges. Just one example was of Moore, 1973. I had temporarily misplaced my personal copy, which I had received in 1973 with the author's complements, when it was published. A used copy cost £ 0.67, but packing and forwarding charges came to £ 8.85! In the intellectually void area where I reside, I could not have hoped to even borrow one from a friend or colleague. After all, how few people the world would even maintain a durable interest in palm classification. There may not even be one in the country where I live!

At the same time, I received enormous help and support from a large number of friends and colleagues, several of them unknown to me personally.

The two librarians of Central Plantation Crops Research Institute (CPCRI) Kasaragod, Mr. Ramesh and Ms. Shobha, were always overly generous with their help and support in procuring and supplying publications. The same should be said of my old friends and colleagues, placed as they are in better-endowed institutions, than I am, M. Ahmedullah, H. Harries, K. N. Nair, and Rajendra Singh. The others, who helped me with publications, photographs, and/or information were W. K. Baker, P. Bellwood, M. Chowdappa, J. Dransfield, J. N. Fenner, B. A. Jerard, P.V. Kirch, L. Noblick, V. Niral, K.M. Olsen, and K. Samsudeen. "My sincerest apologies if i have missed any names." I could not have completed this study without their help.

Now, here are some observations about the contents of the volume. I have included a facsimile of page 1188 of Genera Plantarum, erecting *Cocos* as a new genus and the four figures on coconut ("thenga") in Hortus Malabaricus as the holotype in Chapter 3.

The author will be grateful to receive all the comments and criticisms about the volume.

N.M. Nayar
Emeritus Scientist
Tropical Botanic Garden and Research Institute
Pacha-Palode–695,562
Trivandrum, Kerala, India
nayar.nm@gmail.com
nayar_nm@yahoo.co.in

CHAPTER 1

The Coconut in the World

1. INTRODUCTION

The coconut palm, brings to our mind the fondest recollections of holidays spent on the sunny, sandy beaches of the tropics fringed with swaying coconut palms, especially to those living in temperate regions of the world. It is usually the dominant component of the strand beaches and islands of the tropics in the Indian and Pacific Ocean islands and the Neotropical islands in the Atlantic Ocean. They occur naturally, or are grown in lowlands on the thousands of islands that dot the tropical and subtropical seas of the Indian and Pacific Oceans. They are grown also in the lowlands and occasionally midlands—in the rain-favored or irrigated coastal regions—of much of lowland tropical Asia. For the last 500 years, the coconut has been continuously cultivated in the lowland tropics of West Africa and the Neotropics, where soil and climatic conditions are favorable for its growth.

Tomlinson (2006), the doyen of palm biology, has observed about the uniqueness of palms: "Palms are emblematic organisms of the tropics. They are the world's longest lived trees, because stem cells of several kinds remain active in differentiated tissues throughout the life of the palms. Palms are distinctive from other groups of organisms in that they can make tall and long-lived trees entirely by primary developmental processes, i.e., all the tissues are the direct result of continuously active root and shoot apical meristems."

John Dransfield et al. (2008), an authority on palm taxonomy, have observed that "the often slanting stems and graceful crowns of the coconut are largely responsible for palms being considered the hallmark of the tropics. Furthermore, the coconut, one of the ten most important crop trees, is the mainstay of many people."

The coconut palm is considered the most useful tree to humans. In addition, it used to be literally so for the better part of the 20th century to large numbers of people of the lowland tropics of South Asia and the Indian and Pacific Ocean islands. From about this time, the people of this region, as in the rest of the world, began to pass through a steady transformation in their lives and lifestyles. Every part of the palm until then was being put to active economic uses by the humans in this region (Adkins, Foale, & Harries,

The Coconut
ISBN 978-0-12-809778-6
http://dx.doi.org/10.1016/B978-0-12-809778-6.00001-2

Table 1.1 Production of vegetable oils in the world, 1961[a] and 2011[a]

	Production, 10^6 t	
Item	2011	1961
Coconut oil	**3.1 (9)[a]: 1.95%**	**1.6 (4)[a]: 8.60%**
Cottonseed oil	5.2 (7)	2.2 (3)
Groundnut oil	5.7 (6)	2.5 (2)
Maize germ oil	2.3 (10)	1.5 (6)
Olive oil	3.6 (8)	1.3 (7)
Palm oil	48.5 (1)	0.4 (9)
Palm kernel oil	6.0 (5)	1.1 (8)
Rape and mustard oil	22.9 (3)	0.1 (12)
Rice bran oil	1.1 (11)	0.4 (10)
Sesame oil	1.1 (11)	0.4 (10)
Soybean oil	41.9 (2)	3.0 (1)
Sunflower oil	13.4 (4)	1.9 (5)
Oil crops, others	16.1 (NA)	2.2 (NA)
Vegetable oils, total	**159.2**	**18.6**

[a]Ranking is given in parenthesis.
Source: FAOSTAT, downloaded July 11, 2014.

2002; Foale, 2003; Foale & Harries, 2011; Ohler, 1984). The coconut oil used to be the most-traded vegetable oil in the world till about the 1950s (Purseglove, 1985). Even in 1961, when the Food and Agriculture Organization of the United Nations (FAO) began publishing area–production figures of major crops, coconut oil had ranked fourth (with 9% of total vegetable oil production) among the top 12 vegetable oils of the world (Table 1.1). Now, its position has gone down to a low ninth position for various reasons (with just 2% production of all vegetable oils). The three top positions are now held by palm oil (30%), soybean oil (26%), and rape and mustard oil (canola oil) (14%). During the last 50 years, the global production of total vegetable oil had gone up almost nine times, whereas that of coconut oil, by only two times. Incidentally, as many as 9 out of the 12 major oil crops are annual plants. Only the oil palm, coconuts, and olive are perennial in nature.

2. CONTRIBUTIONS OF COCONUTS TO THE FOOD ECONOMY

2.1 Present Global Production Trends

The coconut is presently grown, or it occurs naturally, in 93 out of 243 countries/territories for which FAO publishes area–production figures

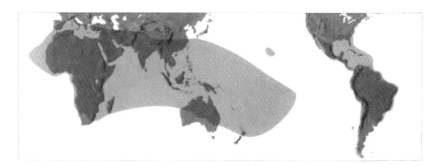

Figure 1.1 Present distribution of palms. *Reproduced with permission from Dransfield, J., Uhl, N. W., Asmussen, C. B., Baker, W. J., Harley, M. M., Lewis, C. E. (2008).* Genera palmarum. *The evolution and classification of palms (732 pp.). Kew, UK: Kew Publishing.*

Table 1.2 Area and production of coconuts, total oil seeds, and rice, 1961 and 2011

Commodity	Particulars	Year 1961	2011	Increase (%)
Coconuts	**Area (M ha)**	5.3	12.0	251.2
	Production (M t)	22.9	57.2	249.8
Oil crops, total	Area (M ha)	113.6	280.1	246.6
	Production (M t)	104.8	550.9	525.7
Rice[a]	Area (M ha)	115.3	215.7	187.1
	Production (M t)	162.3	738.2	455.5

[a]The Rice data are given for comparison.
Source: FAOSTAT, downloaded July 11, 2014.

(FAO Statistics Division (FAOSTAT), 2013) (Fig. 1.1). It is presently (2011) grown in 12.0 M ha area as compared to the total area under oil crops of 280.1 M ha (Table 1.2). This comes to about 4.3% of the area under oil seeds, total. During the 50-years period, 1961–2011, both the area and production in the coconut increased at a comparable pace (+251.2% area, +249.8% production). The increase in area under the coconut also kept about the same pace with other oil crops (+246.6% in oilseeds and +251.2% in coconuts), but, whereas the increase in production of total oilseeds increased by a high 525.7% in the 50-years period, that in the coconut went up by only +249.8%. This more than twofold increase in production in total oilseeds, over that in coconuts, became possible with substantial increases in both yield and production in oilseeds (Table 1.2). For comparison, the respective figures for rice—the most important food crop of humans—for the same period were 187.0% and 454.8%, respectively. The very high increase in total oil-crops production has been achieved mainly by the increases attained in the three oil-crop plants, oil palm (+1340%), rape and mustard (+1666%), and soybean (+994%) (FAOSTAT, Food Balance Sheets (FBS), 2011).

2.2 The Declining Trends in Coconut and Coconut Oil Uses

A comparable declining trend in the relative importance of coconuts as a food item, vis-à-vis total oil crops, is seen also in the food supply of coconut and coconut oil between 1961 and 2011 (Tables 1.3 and 1.4). During this period, their consumption in terms of calories supplied from the increase in consumption of both the items came down marginally from 20 kcal per capita per day during the 50-year period, whereas the consumption in the case of total oilseeds and total vegetable oils more than doubled from 151 kcal (38 + 113 kcal) to 338 kcal (57 + 280 kcal) during the same period (Table 1.4).

Table 1.3 Changes in production and use of coconut and coconut oil, 1961 and 2011

	1961			2011		
	Production 10^6 t	Food %	Processing %	Production 10^6 t	Food %	Processing %
Coconut	22.9	31.0	62.8	57.2	35.8	40.1
Oil crops, total	104.8	14.4	70.6	550.9	8.7	76.7
Coconut oil	1.6	88.8	0.12	3.1	67.8	0.42
Vegetable oils, total	18.6	77.1	0.35	159.2	50.6	0.28

Source: FAOSTAT, downloaded July 11, 2014.

Table 1.4 Changes in food supply of coconuts and coconut oil, 1961 and 2011[a]

	Food supply/capita			
	kg/year		kcal/day	
Item	1961	2011	1961	2011
Coconuts, copra	2.3	3.0	9	11
Oil crops, total	4.9	7	38	57
Coconut oil	0.5	0.3	11	7
Vegetable oils, total	4.7	11.7	113	280

[a]Domestic supply is used for calculations, and not production.
Source: FAOSTAT, downloaded July 11, 2014.

2.3 Causes of Decline in Coconut Use

We shall now look at the reasons for this substantial increase in oil-crops production and use, and the decline in coconut use. In rape and mustard oil, palm oil, and soybean, the increases have been achieved partly by substantial increases in area under cultivation in Indonesia and Malaysia in the case of oil palm, and in Argentina in soybean. The adoption of improved production practices also facilitated this process. Most importantly, however, this was triggered after scientists were able to overcome the presence of antinutritional factors in rape and mustard oil and palm oil (and in cottonseed oil). The major reason for the decline of the coconut and coconut oil appears to be the stigma attached to coconut oil due to the presence of long-chain fatty acids in it. A technological intervention should enable the overcoming of this factor. A decline in the use of coconut oil for industrial purposes has resulted also in the withdrawal of corporates in commercial coconut farming.

2.4 Consumption Patterns

Certain other aspects/factors exist, also concerning the coconuts and coconut oil, that we may look into. During the 50-year period, 1961–2011, whereas the per capita consumption of coconuts registered a marginal increase from 2.3 to 3.0 kg/head/year, that of coconut oil (Table 1.4) went down by as much as 40% to 0.3 kg/head/year. At the same time, the total consumption of all the vegetables oil had increased nearly 2.5 times, from 4.7 kg/head/year to 11.7 kg/head/year.

A significant aside is that the consumption of coconuts continues to be high in several Pacific and Indian Ocean countries as compared to other regions of the world (Table 1.5). It is highest in Micronesia, with a high 121.2 kg/head/year. It is relatively high also in southeast Asia (10.8 kg head/year), the Caribbean (6.1 kg head/year), and South America (7.2 kg head/year) (Table 1.5) as compared to rest of the world. The countries of the world where the consumption of coconuts is highest are listed in Table 1.6. At the same time, their consumption of coconut oil did not rise correspondingly.

The major coconut-producing countries have to be necessarily large countries by area (Table 1.7). In this context, the Philippines, Indonesia, India, Tanzania, and Sri Lanka have the maximum area under the crop in area under cultivation. In production, the leading countries are Indonesia, the Philippines, India, Brazil, and Sri Lanka (Table 1.7). However, these data

Table 1.5 Major coconut growing regions of the world[a] and their coconut and coconut oil utilization patterns, 2011

Region	Production 10⁶/t	Quantity used for[a] Feed %	Processing %	Food %	Food supply/capita kg/year	kcal/day
South Asia						
Coconut, copra	12.4	0.4	24.3	66.5	4.8	20
Coconut oil	0.46	0	0	96.3	0.3	6
Southeast Asia						
Coconut, copra	36.1	0.1	47.5	18.0	10.8	36
Coconut oil	2.33	0	0	31.6	1.2	30
Melanesia						
Coconut, copra	1.0	0	19.1	13.7	70.6	214
Coconut oil	0.02	0	0	18.2	2.12	53
Micronesia						
Coconut, copra	0.16	0	17.1	7.31	121.2	621
Coconut oil	0.004	0	0	0	0	120
Polynesia						
Coconut, copra	0.26	0	52.7	16	91.7	276
Coconut oil	0.0011	0	0	9.1	2.2	38
Caribbean						
Coconut, copra	0.6	0	27.5	37.9	6.1	22
Coconut oil	0.02	0	0	110.0	0.6	15
South America						
Coconut, copra	3.3	0	5.1	89.6	7.2	27
Coconut oil	0.01	0	0	110.0	0	1
World						
Coconut, copra	57.2	0.1	40.9	35.9	3.0	11
Coconut oil	3.1	0	0.4	67.8	0.3	7
Oil crops, total	550.9	6.5	76.7	8.8	7.0	57
Vegetable oils, total	159.2	0.5	0.3	8.1	11.7	280

[a]As given in FAO records.
Source: FAO. FAOSTAT, Basic data downloaded July 11, 2014.

Table 1.6 Countries of the world where the food supply of coconuts and coconut oil is among the highest[a], 2011

	Food supply			
	Coconuts		Coconut oil	
Country	kg/cap/ year	kcal/cap/ day	kg/cap/ year	kcal/cap/ day
1. Fiji	62.9	190[a]	3.2	77[a]
2. Kiribati	123.2	621	5.0	120
3. Philippines	3.4	10	3.4	82
4. Samoa	173.8	530	3.7	91
5. Sao Tome and Principe	136.7	348	1.4	34
6. Solomon Islands	74.3	226	0.8	22
7. Sri Lanka	66.3	272	2.2	55
8. Vanuatu	136.4	374	3.7	50
9. World	3.0	11	0.3	7

[a]Global ranking given in parenthesis.
Source: FAOSTAT, downloaded July 11, 2014.

Table 1.7 Leading coconut-producing countries of the world, 2013

Country (ranking in parenthesis)	Area in M ha	Area in % of net cropped area	Production M t	Yield (t/ha)
Brazil	0.26 (6)[a]	0.35 (11)[a]	2.82 (4)[a]	11.0 (1)[a]
India	2.16 (3)	1.38 (9)	11.93 (3)	5.5 (5)
Indonesia	3.00 (2)	12.77 (3)	18.30 (1)	6.1 (4)
Malaysia	0.11 (11)	3.67 (7)	0.61 (10)	5.4 (7)
Mexico	0.17 (9)	8.50 (4)	1.10 (8)	6.5 (3)
Philippines	3.55 (1)	62.02 (2)	15.35 (2)	4.3 (10)
Sri Lanka	0.42 (5)	8.00 (5)	2.20 (5)	5.2 (8)
Thailand	0.21 (8)	1.26 (10)	1.01 (9)	4.8 (9)
Tanzania	0.68 (4)	4.69 (6)	0.58 (11)	0.9 (12)
Viet Nam	0.14 (10)	2.14 (8)	1.31 (6)	9.6 (2)
Papua New Guinea	0.22 (7)	73.44 (1)	1.20 (7)	5.5 (6)

[a]Figures in parenthesis indicate relative ranking.
Source: FAOSTAT, downloaded October 21, 2014.

do not reflect necessarily the importance of coconuts in the economy and lives of the people of all these countries.

One way of determining this aspect of importance may be to examine the area under the crop in a country as a percentage of its arable area (Tables 1.7 and 1.8). In this manner, among the major coconut-growing countries, coconut ranks first in importance in Papua New Guinea (with 73.3% arable

Table 1.8 Coconut-growing countries of the Pacific and Indian Oceans having relatively large areas under coconuts, 2013

Country	Country area (000 ha)	Arable area (000 ha)	Coconut area (000 ha)	Percent area under coconuts
Fiji	1827.0	165.0	65.0	39.4
French Polynesia	400.0	2.5	22.0	880.0
Kiribati	81.0	2.0	30.0	1500.0
Maldives	30.0	3.0	1.1	36.7
Marshall Islands	18.0	2.0	6.5	325.0
Micronesia	70.0	2.0	17.0	850.0
Papua New Guinea	46,284.0	300.0	220.0	73.3
Samoa	284.0	8.0	27.0	337.5
Solomon Islands	2890.0	19.0	53.0	278.9
Sri Lanka	6561.0	1250.0	420.0	33.6
Tonga	75.0	16.0	9.3	58.1
Vanuatu	1219.0	20.0	98.0	490.0

Source: FAOSTAT, Basic data downloaded October 21, 2014.

land under the crop, but seventh by area and production in the world). This is followed by the Philippines (62.0% area, ranked first by area and second in production in the world), Indonesia (12.8% area; second in area and first in production in the world), and Sri Lanka (8% area, and fifth in both area and production in the world) (Table 1.8).

In terms of supply of coconut, the present situation in the world as a whole is dismal. It is even worse in the case of coconut oil (Table 1.6). The present daily per capita consumption (as food) of coconut and coconut oil is only 8.2 g/head/day of coconut and 0.8 g/head/day of coconut oil.

A very significant situation is that seven countries in the world—all in the Pacific Ocean—have much higher areas under the coconut than their respective arable land area (Table 1.8). This ranges from 279% in Solomon Islands to as high as 1500% in Kiribati, 880% high in French Polynesia, and so on. This can be explained only by assuming that the area under the coconut over and above that of the total arable land area represents area under self-sown and naturally occurring coconuts. This group of islands, as well as those highlighted in Table 1.8 emphasize the continuing very high dependence of most island nations in the Pacific and Indian Oceans (Maldives, Sri Lanka) on the coconuts. A comparable situation does not seem to exist in the case of any other crop plant.

CHAPTER 2

Early History, Lore, and Economic Botany

1. INTRODUCTION

Presently, coconuts are found in the tropics and subtropics in practically all of the thousands of islands that dot the Pacific and Indian Oceans and the West Indies and, to a lesser extent, along the coastal lowlands of South and Southeast Asia and parts of east Africa, where even soil and climatic conditions are favorable for its growth. This has been so from prehistoric times in most of the Asia-Pacific region. On the American continent, on its west coast, coconuts, or other palms looking broadly similar to the coconuts, were observed by early Spanish navigators, in two to three locations including the Panama coast after Columbus' landing in the West Indies (Hispaniola) in 1492 CE. In the Atlantic Ocean, the Portuguese had carried out most of the introductions following Vasco da Gama's landing in southwest India (Calicut) in 1498 CE and Pedro Alvarez Cabral's discovery of Brazil and subsequent landing in Calicut, Kochi, and Goa in 1500 CE.

Unlike the Chinese and Mediterranean cultures, the written script came late in the Pacific and Indian Ocean societies. In the Pacific Ocean societies, it came only after the arrival of the Europeans in the region. At the same time, in South Asia, the rich oral accounts contain numerous references about the use of coconuts and coconut's plant parts in numerous religious and social contexts. In South and Southeast Asia, writing was developed comparatively late, in only the late second millennium BCE. Early texts of the period contain many references on the use of coconuts in numerous religious and cultural traditions. We shall now undertake a brief overview of them.

2. ETHNOBOTANY OF PALMS

It may be correct to state that Arecaceae is the only plant family that would have been able to sustain all the basic needs of present-day humans till the start of the Anthropocene. This observation is attested from the ethnobotanical studies of palms reported from Central and South America, and the

The Coconut
ISBN 978-0-12-809778-6
http://dx.doi.org/10.1016/B978-0-12-809778-6.00002-4

Pacific Ocean islands (Balick, 1984; Camara-Leret, Paniagua-Zambrana, Balslev, & Macia, 2014; Johnson, 2010; Moore, 1973). The following account has been taken mainly from the first two publications. The record is meant to be illustrative and not exhaustive.

Balick (1984) had quoted from a report of the great naturalist, Alexander von Humboldt (1853) that an Amazon basin group, the Guaraon Indians, were then meeting all their requirements from a single palm *Mauritia flexuosa* L. (Tribe *Lepidocaryeae*). A limited search conducted by this author did not show the existence of such a group. But it showed the presence of two language groups, Guarani and Guarijio. *Mauritia* is a massive, stately solitary palm widely present in South America and possessing robust erect stems and huge palmate leaves.

Dransfield et al. (2008) have recorded that the palm is immensely useful as a source of oil, starch, wine, timber, cork, and fiber for weaving and tying, and palm hearts. From the accounts given by various authors, we can possibly designate *Mauritia* as a keystone species.

Moore (1973) had observed that palms were being used in the greatest number of ways by the American Indians. The Neotropics and the Pacific Ocean islands are home to the largest number of palm species. Their uses included making houses, baskets, mats, hammocks, crates, quivers, pack baskets, impromptu shelters, blow pipes, bows, starch, wine, protein from insect larvae reared within trunks or in the exudates, fruits, beverages, flour, oil, ornaments, loincloths, cassava graters, medicines, perfumes, and so on.

The three palm species that are most used in the Neotropics have been *Oenocarpus bataua* in the Amazon, *Attalea phalerata* in the Andes, and *Iriartea deltoides* in the Choco region (Camera-Leret et al., 2014).

Johnson (2010) had observed that the coconut was being put to 30 uses in various regions of the world, especially in the Pacific Ocean islands and South and Southeast Asia. Johnson observed that Trukese people living in the Truk Islands (Caroline Islands, 1100 km southeast of Guam) used the coconut palm for the largest variety of purposes.

3. PACIFIC OCEAN ISLANDS
3.1 General
The coconut has had the most continuous development here.

The Pacific Ocean is the largest of the three major oceans of the world. The other two are the Indian and Atlantic Oceans. It has a spread of over 165 M km^2 water area and 1.3 M km^2 of land area comprising about 25,000

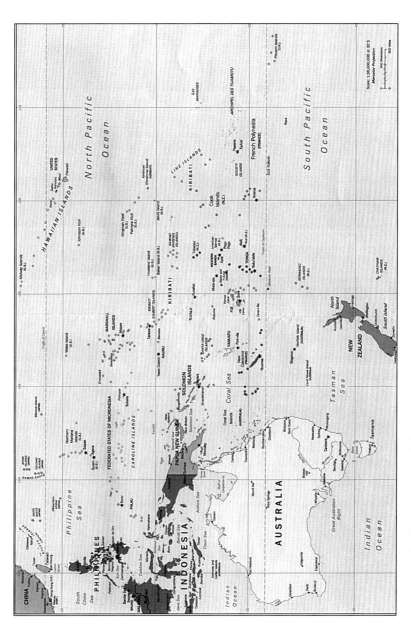

Figure 2.1 Pacific Ocean political showing places mentioned in the text. *Source: lib.utexas.edu (Perry Castaneda Library).*

islands (see Fig. 2.1). The population is about 1.2 M people living in just about 800 islands only and speaking about 1200 languages. The islands are collectively known as Oceania (Wikipedia, 2015).

The Pacific Ocean measures 16,640 km at the equator (Wikipedia, 2015). Most islands enjoy a tropical climate even though a few island groups extend beyond the tropics into the subtropics. Broadly, there are two kinds of islands, the high volcanic islands, and the generally small and low atolls. Papua New Guinea (PNG) (0.8 M km² area), the second largest island in the world, is the western limit of the Pacific Ocean. Oceania is traditionally identified into three regions, Micronesia, Melanesia, and Polynesia. More recently, to avoid racial connotations, the Pacific Ocean Islands are grouped as Near and Remote Oceania (Fig. 2.1). Oceanians have been noted for their high voyaging skills and rich oral traditions (Bellwood, 1987; Kirch, 2000).

3.2 Peopling the Pacific: The Role of the Coconut

The modern day humans (*Homo sapiens*) had first populated New Guinea and Australia c.50,000 ya (years ago). The rest of the region toward the east remained uninhabited till the late Holocene period [till after c.3000 BCE (before the Common Era)] (Bellwood, 1987, 2013; Kirch, 2000).

The hundreds of islands dotting the western Pacific Ocean and lying to the east and southeast of the Asian mainland (covering the present Philippines, Indonesia, parts of Malaysia, and Taiwan, and lying to the west and north of New Guinea), were first populated by modern-day humans c.67,000 ya (Bellwood, 1987, 2004, 2013). Earlier, they had migrated south from the lower Huang He river valley in present-day northeast China. There, they developed new customs and came to be called Austranesians. From there, they moved to south China, then to Taiwan, and from there, they crossed the sea and reached Luzon (Philippines). Another group traveled south to the present continental Malaysia, and then on to the present Indonesia. These people were using initially watercraft, grinding stones, and edge-ground pebbles. They subsisted on littoral resources, hunted in coastal rain forests, and crossed seas up to 150-km width to reach Australia and Papua New Guinea (Bellwood, 2013).

Incidentally, when the Austranesians arrived in the Pacific Ocean islands, the coconut was a component—a major one in some locations—of the already existing vegetation. This has been attested by archaeological findings (Bellwood, Fox, & Tryon, 1995).

Further expansion of humans appeared to have taken place after only the development of agricultural economies, which happened after 4000 BCE (Bellwood, 1987, 2013). The colonization of Polynesia (other than New

Guinea) began in Mariana Islands about 1500 BCE. These Neolithic people colonized Melanesia during 1350–900 BCE. These Lapita people carried out their cooking in earthen ovens and open hearths. They also invented double sailing canoes. They developed terrace farming of taro in New Guinea using canal irrigation (Bellwood, 2013).

Several authors have pointed out that east of the Bismarks (PNG), there were only a few plants having edible fruits. The most common ones were *Pandanus*, *Barringtonia* spp., *Cocos nucifera*, *Rhizophora*, and *Metroxylon* spp. (Bellwood, 2013; Kirch, 2000).

There was a long pause of about 1600 years after the colonization of Tonga and Samoa during 900−800 BCE in settling the Pacific islands (Bellwood, 2013; Kirch, 2000). Thus, central and southeast Polynesia (Cook, Austral, Mangarewa, Pitcairn, etc.) were populated by only 900–1000 CE and the Society Islands (17°00′S 151°00′W) by 800–1200 CE. The last of the Pacific Ocean islands were occupied by present day humans (*Homo sapiens*) only in the last 1000 years.

4. ECONOMIC BOTANY OF THE COCONUT

The Austranesians had no written script, but they possessed rich oral traditions. Hence, all the information that we have of Polynesian history and of food-production systems have come from the recordings made by some early settlers after the European occupation (which began from early 16th century only) and from the accounts of a handful of early navigators following Magellan's first Pacific crossing in 1520 CE. Both archaeology and palynology have contributed to this knowledge.

We had noted in the last section about the colonization of the Pacific islands, the diverse time of settlement of the people in different regions, and the extended time line of their occupation. This appears to have contributed to the variability in the uses of the locally available natural resources of the region by the people.

Thaman (1992) reviewed over 120 publications relating to the uses of plants by the people of Oceania (also, Prigge, Lagenberger, & Martin, 2005). The following account has been summarized from this work, unless mentioned otherwise.

A total of 140 species of strand vegetation were being used by the people for various purposes. This consisted of 10 ferns, 17 herbs, 11 grasses and sedges, 14 vines and lianas, 26 shrubs, and 62 trees. This list excluded the plants that were being used in only a few islands and also those confined to inland forests. The 140 species were being used for 75 different purposes/

use categories. The total frequency of the use was 1024, ranging from nil (in two species) to as many as 125 uses (in the coconuts). This was followed by 17 species with 20–57 uses, and another 29 species with 7 uses each. Almost a quarter (actually, 27 species) were being used for medicinal purposes.

The medicinal use of the coconut was one of its 125 uses. The coconut was also one of 16 species used in general construction. All parts of the coconuts were being used as fuel, especially in the atolls. The shell was the most preferred item in the coconut. Boat sails used to be made with plaited leaves of either the coconut or *Pandanus*. The coconut was the most favored plant for making musical instruments. Several legends and songs have been in circulation based on the coconuts. It was also one among the six most important staple food plants of the Pacific. The others were sago palm (*Metroxylon*), Polynesian arrow root (*Tacca leontopetaloides*), mature seeds of *Pandanus tectorius* and *P. pulposes* (and their tender roots), and mature seeds of the Tahitian chestnut, *Inocarpus fagifer*. Thus, until recently (till the end of the Holocene and the start of the Anthopocene, shall we say!), the coconut was the most useful of all plants in the Pacific until recently. The well-known French botanists of the Pacific, Massal and Barrau (1956), had observed that "human life on atolls would be scarcely possible without coconuts." They had stated that in some atolls, the daily per capita consumption of the coconut used to be up to six nuts (but see Tables 1.7–1.9) (refer also, Abe & Ohtani, 2013; Barrau, 1965).

Throughout the Pacific, the coconut features in the mythologies, legends, songs, proverbs, and riddles of the peoples. It was of much ceremonial importance in Polynesia and Micronesia displaying its leaves used to be a sign of higher rank. In Tuvalu, the leaf tip of the coconuts was being used as a religious emblem. Thus, we can see that coconut was literally the staff of life for the peoples of Oceania (Balick, 2009; McKillop, 1996).

In addition to the aforementioned accounts, there are also reports from the Indian and Pacific Ocean regions about the cultural and medicinal uses of the coconut.

Krauss (1974) studied the ethnobotany of Hawaii Islands. She observed that there was no clear temporal evidence about the first arrival of permanent settlers to the Islands. However, it was widely believed that the first residents came possibly by 500 CE from either Marquesas or Tahiti, and certainly before 750 CE. The six major crops of the Islands have been coconuts, taro, breadfruit, sweet potato, paper mulberry (*Brouessonetia papyrifera*, Moraceae), and candlenut (*Aleurites moluccana*, Euphorbiaceae).

Every part of the coconut palm was being put to active use: the swollen trunk base used for making drums, the trunk for making hull canoes and buildings, the petiole base as pounder, the fibrous sheath for food packing, the full leaf used as a symbol of high rank, midribs of leaves for making brooms and candles, and fiber for making sennit. However, the plaited leaf was not being used in Hawaii, and the use of coconut milk appeared to have commenced there only after the visit of Captain Cook.

Kitalong et al. (2011) studied the uses of local plants in Palau archipelago, western Micronesia (7°30′N 134°30′E) with emphasis on their medicinal uses for four seasons (2006–2010 CE). They recorded 448 different uses involving 170 plants. They were grouped into 28 categories. This included 80 species used for 235 medicinal uses. The most widely cited species was again the coconut. It was being used in medicine, as food, and in construction, the last being the most important use. Coconut oil, too, was being used for diverse purposes—in cooking, as emollient, for ceremonial bathing of newborn babies, and for the "purification" ceremonies of the mother after childbirth and after the first menstruation. The consumption of coconut oil was supposed to strengthen the immune system in humans. The coconut had 39 separate uses followed by the arecanut. The authors designated these two plants, coconut and arecanut, as the "cultural keystone species" in Palau.

Bourdeix, Johnson, Tuia, Kape, and Planes (2013) have given a rambling account of the seemingly very close relationship of the people of French Polynesia with the coconut. Firstly, the locals believe that the coconut could be classified into male and female palms. The female palms are preferred over male palms for planting. The male and female palms are recognized by four attributes: (1) If the distal end of the nuts shows a depression or cavity, then it is a female. The rest are considered male. (2) If new sprouts in a seed nut emerge from the peduncle end of the fruit, then it is a female, if from any other part of the nut (fruit), it is a male. (3) If the shape of the newly emerged leaf from a fruit is a bit broad, it is considered a female; if they are narrow-leaved, they are male. (4) The heavy-yielding palms are female and the rest are male. The author was apparently unable to determine any rationale for this belief.

Bourdeix et al. felt that, in general, the local people possess only a low level of technical knowledge of the coconut palm. Still, they get very offended—actually angry—if anyone inflicted any cut or damage to the coconut palm. This often included also any suggestions to harvest tender nuts for drinking its water. The authors did not indicate how this belief could have synced with the widely held assumption that in the early period of colonization of Oceania, the Polynesians used to routinely carry tender coconuts as the source of drinking water, especially because small islands and small atolls do not usually have any sources of freshwater.

Bourdeix et al. have also mentioned the presence of certain rare forms of coconut in Polynesia. They are listed here: (1) Niu utongau variety (Onike island, Haapai group, Tonga) and a landrace from Cook islands, niu mangarao, possess sweet husk. (2) A landrace, Rennell Tall, found in Rennell (Solomon islands) possesses one of the biggest nuts observed in the coconut. The author did not give any further details about any of them.

5. INDIAN OCEAN

5.1 General

The Indian Ocean is the smallest in size and the youngest in age among the three major oceans of the world with $73 \, M \, km^2$ area and 20% of the water area. It arose only about 150 Ma following the Gondwana split. It is bounded in the north by Asia and the west by Africa. It merges with the Pacific Ocean in the region north of Australia and with the Southern Ocean in the south. There are about 65 island groups, but the total number has not been determined properly. It may be about 8000 islands. The ocean is about 10,000-km wide between the southern tip of Africa and northern tip of Australia. Compared to the Pacific Ocean, there are fewer islands and archipelagoes in this ocean (Fig. 2.2). The two major islands are Madagascar ($0.59 \, M \, km^2$ area; the fourth largest in the world) and Sri Lanka (Wikipedia, 2015).

The coconut is an essential component of the lowland coastal vegetation in much of the mainland of South Asia and in most of the hundreds of islands of the Indian Ocean (Sauer, 1967).

5.2 The Coconut in Social and Religious Context

The people of South and Southeast Asia follow mainly three religions, Buddhism, Hinduism, and Islam, and their variations, besides a few minor ones. Hinduism and Buddhism are replete with references and symbolisms involving the coconut (Ahuja, Ahuja, & Ahuja, 2014; Gupta, 1996; Gandhi & Singh, 1989; Possehl, 1997; Randhawa, 1964). Unlike the civilizations in China and the Mediterranean region, written script was developed in South Asia only over three millennia ago (Thapar, 2002). Hence, till then, all the ancient scriptures and epics, such as the Vedās, Purānas, Ramāyana, Mahābharata, etc., were being maintained through memory (oral traditions) only; they were converted to textual form only in the early first millennium CE (Singh, 2009).

In the Hindu and early Buddhist and Jain religions, the coconut is used in several customs and rituals.

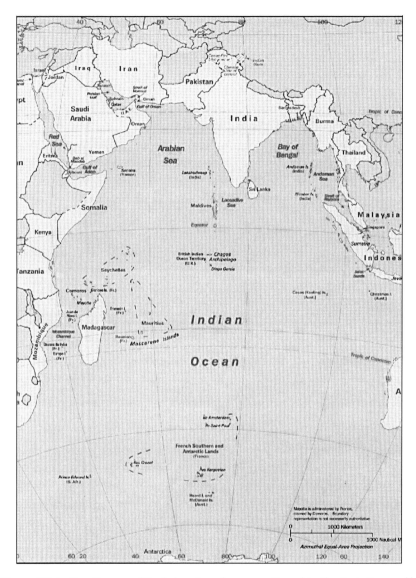

Figure 2.2 Indian Ocean political showing places mentioned in the text. *Source: lib. utexas.edu (Perry Castaneda Library).*

In south India, the husked nut has been used since ancient times in many religious and social rites and rituals, as also the unopened inflorescence and tender leaves. Most of these practices continue to this day among the Hindus, including in the interior regions of India, where the coconut is not grown.

A most-prevalent practice is to break a coconut before the god Ganesha at the start of an event to remove all the obstacles and to invoke the blessings of the Almighty. This is done even on such occasions as the launch of ocean-going ships. The tender leaves are used for decorating the premises for many social and religious functions, including weddings, inauguration programs, etc. Likewise, during Hindu weddings, the unopened inflorescence is taken out from the spadix and put up in a special wooden vessel (called *para*, which is used for measuring paddy; c. 16 kg capacity) filled with paddy. It is meant to invoke prosperity and be a witness to the solemn event.

Then on occasions, when celebrated or distinguished persons are to be received formally, a purna kumbham is used (Fig. 2.3) in the ceremony. It consists of a pitcher, usually made of silver or copper (also brass or gold) filled with water and topped with a husked coconut and with 5, 7, or 11 mango leaves surrounding the coconut (Fig. 2.3A and B). Purna kumbhams are depicted in several south Indian temples and the central Indian Buddhist and Jain temples of Sanchi, Madhya Pradesh, in central India (Gupta, 1996; Randhawa, 1964).

Such and other customs are most prevalent in the Kerala and Konkan coasts where the coconut abounds. Some of these customs have been adopted even in certain Christian rites and festivals of the region. These customs are widely observed in peninsular India and to a lesser extent in southeast and east India.

5.3 Origin of the Coconut

The Vishnu Purāna (c.1200–600 BCE) gives the following story on the origin of the coconut (cf. Thapar, 2002). Once, the Sage Viswāmitra wanted to practise penance in a distant land for an extended period. The Sage had then left his family behind. A severe famine broke out in the region while he was away. The king of the region, Satyavrata, came to know of the difficult condition of Viswamitrā's family. He then took the family under his care. When the Sage returned home after concluding his penance, he learned about the protection extended to his family by the King. Viswamitra felt very happy. He asked the king to seek any favor. The king requested the Sage to send him (Satyavrata) to the heaven along with his body (normally, only the spirit of the mortals are permitted entry into the heaven). Upon seeing the ascent to the heaven of King Satyavrata with his body, its keeper Indra, got upset and pushed Satyavrata back to the earth. When Viswamitra saw what was happening, he felt insulted and decided not to let Satyavrata fall on the earth. Using his superpowers, he erected a pole and allowed Satyavrata to

Figure 2.3 Purna kumbham/Kalasha. (A and B) Various views, (C) general view. The national head of the cultural organization, Sri Ramakrishnan Mission, being received with a purna kumbham, (D) purna kumbham in stone: from Nageswaram temple, Kumbhakonam, Tamil Nadu (TN), India (9th century CE), (E) purna kumbham motif as pillar decoration: Nataraja temple, Chidambaram, TN, India (16th century CE). *(A and B) Wikipedia (Creative Commons License). (C) Sri Ramakrishna Math, Trissur, Kerala, India. (D) American Institute of Indian Studies, New Delhi. (E) Archaeological Survey of India, New Delhi.*

rest over it. This world came to be known as Trisanku, a new heaven that was "neither here nor there." Gradually, this pole carrying Satyavrata developed into the coconut palm (see Fig. 2.4) (Gupta, 1996; Wilson, 1961) that we have today.

Figure 2.4 Sage Viswamithra creating a coconut palm, in stone. Jambukeswar temple, Thiruchirapalli, TN, India (17th century CE). *Courtesy: Archaeological Survey of India.*

5.4 Other Rituals

The breaking of a coconut before the start of an event is stated to be symbolic of the ancient practice of human sacrifice to propitiate the deity Kāli. Gradually, the sacrifice of the humans was substituted with animals, and later on, by the coconut, so the belief goes. The three "eyes" present on the husked nut are stated to represent a human face (Gupta, 1996).

Gupta (1996) has listed 78 plants that appear in the stone carvings of Hindu, Buddhist, and Jain temples in India. The coconut is one among the most commonly depicted trees. These temples are located mostly in south and central India. Some of the more well known among them are: Thillai Nataraj temple, Chidambaran 11°20′N 79°45′E (originally, early Common Era, the present one 12–13th century CE); Meenākshi temple, Madurai (09°54′N 78°06′E early Common Era); Nageshwara Swami temple, Kumbhakonam (10°59′N 79°23′E; 9th century CE); Viswabrahma and Virabrahma temples, Alampur, Telengana (15°88′N 78°132′E; both 7th century); Khajurāho temple, Madhya Pradesh (28°48′N 79°55′E; 950–1050 CE, Jain and Buddhist temples); and Sanchi Stupa and temple, Madhya Pradesh (23°29′N 77°44′E; 3rd century BCE, Buddhist). Madhya Pradesh is not a coconut-growing area now; however, this is a region where several fossil coconuts have been recovered as part of the Deccan intertrappean biota.

The Roman period had an extended connection with the coconut as the following account will show (Fig. 2.5).

(A)   **(B)**

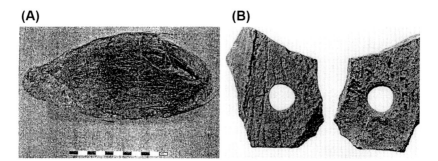

Figure 2.5 (A) Fruit with partly removed husk. (B) A piece of shell, outer and inner views. Remains of coconut recovered from the Roman period port of Berenike, present-day Egypt. *Reproduced from Cappers, R. T. J. (Ed.). (2006).* Roman footprints at Berimike. *Los Angeles, CA, USA: Cotsen Institute of Archaeology, University of California. 229 pp. with permission.*

5.5 The Roman Connection

The Roman empire lasted for 7–8 centuries. It was at the height of power during c.27 BCE−476 CE (Mokhtar, 1981). Spices, textiles, elephant tusk, semiprecious stones, etc. from the Orient, including India, present-day Indonesia, and China, used to be in great demand in the Roman Empire for different uses—in medicine, food, and rituals. This trade was in Arab hands mostly, and also with some Indian tradesmen. The Arab tradesmen used to employ part-overland and part-sea routes to carry out this trade—generally, by sea, up to the Arab or Iran coasts, and to the north Red Sea ports and by land thereafter.

This trade has been well documented in the writings of Pliny, Ptolemy, the Periplus, and so on. The Periplus Maris Erythracei is an anonymous account of the times, trade, and shipping carried out in the Indian Ocean including the Red Sea. This is assumed to have been prepared by a Greek trader in the 1st century CE (Casson, 1989; Schoff, 1912).

The main Roman ports in the Red Sea were Berenike and Myos Hormos in the Red Sea in Egypt.

Archaeological investigations conducted in these ports have given evidence for the presence of 32 botanicals. This included rice, black pepper, and coconut shells (Cappers, 2006; Sidebotham, 2011; van der Veen, 1999). The coconut and rice remains are assumed to have been left behind by the merchants from India, several of whom used to live in these port towns for extended periods (Fig. 2.5).

6. ATLANTIC OCEAN

The coconut has a history of only about 600 years in this region.

During the Middle Ages, India (and the Far East) were known as the land of spices. There was intense competition among the then major European powers to capture the spice trade. Until then, it was the monopoly of Arab merchants who used to take the overland route to Europe from the Middle East. This was after the spices and other items (elephant horn, semiprecious stones) were brought from the west coast of India by coastal shipping using the famed dhows. In their bid to capture the trade, the Portuguese succeeded in discovering the first sea route to India via the Cape of Good Hope. This followed the landing of the Portuguese navigator, Vasco da Gama, in Calicut (in southwest India) in 1498 CE and Cabral in 1500 CE. They had then visited Cochin and Goa as well. The Portuguese then went on to establish several new outposts in South and Southeast Asia.

The Spanish navigators soon followed the Portuguese. They discovered the sea route to the Far East around South America. This resulted in the Spanish adventurer Magellan landing in the Philippines in 1521 CE. All the later Spanish navigators had reached the Philippines by crossing the Pacific Ocean around Cape Horn in South America and across the Atlantic Ocean (see Fig. 2.6).

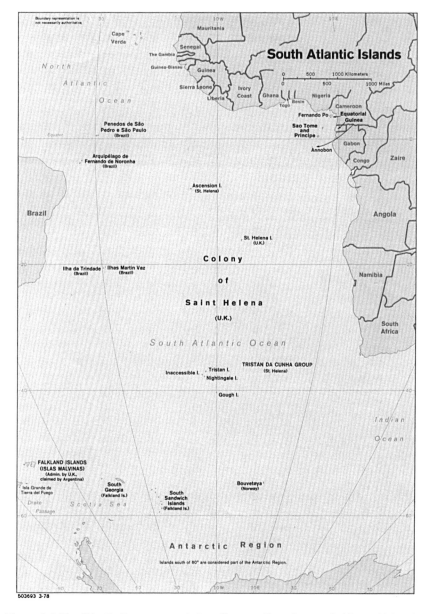

Figure 2.6 The Atlantic Ocean general view. *Courtesy: Perry Castaneda Library, University of Texas, Austin, TX, USA.*

The history of the introduction and spread of coconuts in the Atlantic Ocean is tied to these quests for spices by the then European powers. The Portuguese and Spanish were followed by navigators from several other European countries such as the English, French, and later the Belgians, Italians, and Germans.

Much earlier to the Indo-Roman trade, the Austranesians from mainly Indonesia had reached Madagascar from the early centuries of the Common Era by boat and had begun to settle there. Moreover, they introduced many tropical plants in the region including the coconut and spread their culture (Blench, 2007; Murdoch, 1959; Verin, 1981, etc.).

This account shows that the peoples of South and Southeast Asia and the western Pacific Ocean region were familiar with the coconuts from prehistoric times. The coconut had become an integral part of the cultures and traditions of South Asia from at least one to two millennia before the Common Era. The coconut was incidentally part of the natural vegetation then in these regions (cf., Chapter 7).

CHAPTER 3

Taxonomy and Intraspecific Classification

1. HISTORY OF COCONUT TAXONOMY

The coconut is taxonomically *Cocos nucifera* L. [Sp. Pl.: 1188 (1753), according to the latest classification of the family Arecaceae (Dransfield et al., 2008; termed here GP II, Genera Palmarum II)] (Fig. 3.1).

The coconut palm belongs to the family Arecaceae (also known as Palmae), subfamily Arecoideae (one of five subfamilies), tribe Cocoseae (one of 14 tribes in the subfamily Arecoideae), subtribe Attaleinae (one of three subtribes in the tribe Cocoseae), and genus *Cocos* L. (one of 12 genera in the subtribe Attaleinae). At present, the genus *Cocos* is monospecific, with just one species, *nucifera,* included in it. However, this was not always so, as we shall see presently. The classification of the family Arecaceae has been undergoing major course corrections in all the family-level classifications published from time to time. This is clear from comparison of the past classifications of the family published in GP II (Dransfield et al., 2008, pp. 134–135).

These changes have been attributed to several reasons: inflow of more new information, mainly from phylogenetic studies; gaps in available information, because of relative paucity of studies; generally poor quality (mainly, incompleteness) of herbarium sheets; prevalence of homoplasy, and so on. To illustrate this point, comparisons from the tribe Cocoseae of the latest two classifications are given in Fig. 3.2A and B. Although the classification shown in Fig. 3.2A (Dransfield & Uhl, 1998) was based entirely on the basis of morphological characters, that in Fig. 3.2B (Dransfield et al., 2008) has been based mainly on phylogenetic relationships.

Linnaeus (1753) had used as type specimen the figures of the coconut palm and its parts given by van Rheede Tot Drakenstein (1678–1693) in their publication, Hortus Malabaricus (Fig. 3.1). Because we have not seen this type specimen included in any of the published volumes on the coconut, we have included it here.

As stated before, the present monospecific status of the coconut palm, *Cocos nucifera,* was not so always. Until about 100 years ago, new species were

The Coconut
ISBN 978-0-12-809778-6
http://dx.doi.org/10.1016/B978-0-12-809778-6.00003-6

COCOS

nucifera   1.   COCOS frondibus pinnatis: foliolis replicatis.

Coccus frondibus pinnatis: foliolis enfiformibus margine villofis. *Hort. cliff. 483. Fl. Zeyl. 391. Roy. lugdb.* 4.

Palma indica coccifera angulofa. *Baub. pin.* 502.

Palma indica nucifera. *Baub. bift.* 1. p. 375.

Calappa. *Rumpf.* 1. p. 1. t. 1. 2.

Tenga. *Rheed. mal.* 1. p. 1. t. 1. 2. 3. 4.

Habitat in Indiae *paludofis, umbrofis.*

Foliola omnia (excepto utrinque infimo) retro-plicata

funt, contra ac in fequente.

Figure 3.1 The type specimen of the coconut, *Cocos nucifera* L., along with the description and figures given in Linnaeus (1753). *Reproduced from Linnaeus, C. (1753). Cocos (p. 1188). http://bioversitylibrary.org/page 359 209. Downloaded 02 January 2016 (Figs. 1, 2, 3, and 4).*

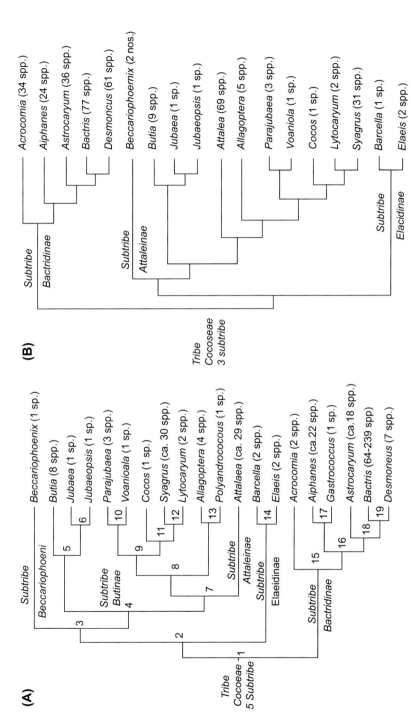

Figure 3.2 Relative position of the genus *Cocos* in the latest two palm family classifications. *(A) Reproduced from Dransfield, J., & Uhl, N. W. (1998). Palmae. In K. Kubitski (Ed.), The families and genera of vascular plants. IV. Flowering plants. Monocotyledons (pp. 306–389). Springer Berlin. (B) Reproduced from Dransfield, J., Uhl, N. W., Asmussen, C. B., Baker, W. J., Harley, M. M., Lewis, C. E. (2008). Genera Palmarum. The evolution and classification of palms (732 pp.). Kew, UK: Kew Publishing.*

being added in the genus *Cocos*. In 1916, there were more than 90 species and varieties in the genus. Beccari (1916) then transferred all of them except *Cocos nucifera* to other genera, mainly to *Syagrus* and *Butia,* and left only one species, *nucifera,* in the genus *Cocos* (Beccari, 1916; Glassman, 1987, 1999). Later palm taxonomists have accepted these changes (e.g., Dransfield et al., 2008; Moore, 1973). Thus, *Cocos* has continued to be a monospecific genus since 1916. The current International Plant Names Index (IPNI) list (2015) gives 180 *Cocos* names, determined as synonyms of *C. nucifera* (www.kew.org/wcsp).

2. CLASSIFICATION OF TRIBE COCOSEAE

The following account has been prepared largely from Dransfield et al. (2008), GP II.

The tribe Cocoseae to which the coconut belongs includes 18 genera and c. 367 species arranged in three subtribes, Bactridinae, Elaeidinae, and Attaleinae. All the species except six (including *Cocos*) are native to Central and South America, including the Caribbean. The subtribes Bactridinae and Elaeidinae are armed and Attaleinae is unarmed.

Tribe Cocoseae: Slender to very robust, acaulescent to erect or climbing habit; leaves pinnate or pinnately ribbed; inflorescence unisexual or bisexual, spicate or branched to one, or very rarely, two orders, bearing a short prophyll; flowers superficial, or occasionally in pits; petals or pistillate flowers imbricate or connate; gynoecium with three, or rarely, more locules and ovules; fruits sometimes very large, one to several seeded, never lobed; endocarp with three or more well-defined pores.

2.1 Subtribe Bactridinae

A very diverse group consisting of five clearly demarcated genera; diminutive to very large, acaulescent, erect or climbing. The genera are *Acrocomia* (c. 34 species, chromosome number 2n = 30), *Aiphanes* (c. 24 species, 2n = 30), *Astrocaryum* (c. 36 species, chromosome number not known), *Bactris* (77 species, 2n = 30), and *Desmoncus* (66 species, 2n = 30); total, c. 237 species; native to Central America, Caribbean, and northern South America.

2.2 Subtribe Elaeidinae

Moderate to large; includes two genera, *Barcella* and *Elaeis,* and three species.

The genus *Bactris* includes a single species. It is found in Lower Amazonia. Its chromosome number is not known. The genus *Elaeis* includes two species, *E. oleifera* and *E. guineensis*. *E. oleifera* is native to tropical West Africa. It is known as the African oil palm of commerce, which is presently the top producer of vegetable oil in the world. Nevertheless, 90% of the oil palm is now cultivated and produced in Southeast Asia, almost entirely in Indonesia and Malaysia. In tropical West Africa, *E. oleifera* occurs naturally in a semidomesticated state (Zeven, 1967). On the other hand, *E. guineensis* is native to Central and South America. The chromosome number of both species is 2n = 32.

2.3 Subtribe Attaleinae

It is the third subtribe in the tribe Cocoseae. It includes 11 genera and c. 125 species. All the species are unarmed.

A. *Beccariophoenix*: Large, solitary, unarmed, pleonanthic, monoecious; endemic to the central and eastern regions of Madagascar; two species; chromosome number, 2n = 36.

B. *Jubaeopsis*: Moderate, clustering, pleonanthic, stems short and erect that frequently branch dichotomously; endemic to southeastern South Africa; occurs in the coastal reaches of two rivers; includes one species; The chromosome number is 2n = 160–200.

C. *Voaniola*: "The forest coconut"; solitary, unarmed, pleonanthic endemic to northeast Madagascar; in primary forests, in swampy valley bottoms, and on gentle slopes up to 400 m; one species. The chromosome number, 2n = 550–600, is the highest in the monocotyledons.

D. *Allagoptera*: Small, acaulescent or moderately erect palms; five species in eastern South America (Brazil and Paraguay); inflorescence spicate 6–100 stamens; in loose sand on beaches and dunes, in open tree and scrub woodlands, on sandstone hills, and in dry grasslands (cerrado); fruits somewhat irregularly shaped because of tight packing on spikes; chromosome number, 2n = 32.

E. *Attalea*: Solitary, small to massive; pinnately leaved, often huge-leaved palms; inflorescence staminate or pistillate, or carry flowers of both sexes on same palm; fruit generally large with very thick endocarp, 1–3 or more seeded; native to Central and South America and the Caribbean (form Mexico, southward to Bolivia and Peru); occurring in a wide range of habitats, from tropical rainforests to dry cerrado; c. 69 species; chromosome number, 2n = 32.

F. *Butia*: Small to moderate, solitary or clustered, pinnate-leaved palms; native to cooler and drier parts of South America (south Brazil, Paraguay, Uruguay, Argentina); in grasslands, cerrado, and woodlands in lowlands; nine species; chromosome number, 2n = 32.

G. *Cocos*: "The often slanting items and graceful crowns of the coconut are largely responsible for being considered the hallmark of the tropics" (Dransfield et al., 2008).

Moderate, solitary, unarmed, pleonanthic, monoecious palm; a single species; widely cultivated throughout the tropics and warmer subtropics; differs from other genera in Attaleinae in having large pistillate flowers, with rounded sepals and petals, in the large fruit with thick fibrous mesocarp, and in the endosperm. Chromosome number 2n = 32.

H. *Jubaea*: The Chilean wine palm; one of the most massive of all palms; native to central Chile; growing on sides of ravines and ridges in dry scrubby woodland; one species; chromosome number, 2n = 32.

I. *Lytocaryum*: Graceful undergrowth palm from southeast Brazil; at 800–1800 m with distinctive slender discolorous leaflets and fruits; slender, solitary, unarmed, pleonanthic, monoecious palm; two closely related species; chromosome number, 2n = 32. Recently, Noblick and Meerow (2015) have merged this genus with *Syagrus*.

J. *Syagrus*: Extremely variable genus, native to the Caribbean Lesser Antilles (one species) and South America (31 species from Venezuela southward to Argentina); most species occur in dry to semidry areas; all acaulescent; a few usually tree-like species are restricted to mesic and tropical rainforests; small to tall, solitary or clustered, rarely forking below ground, unarmed or armed, pleonanthic, monoecious palms. Chromosome number, 2n = 32.

K. *Parajubaea*: Large pinnate-leafed palms of high altitude in inter-Andean valleys in South America; three species, in Ecuador, Bolivia, and Colombia; in high elevations, 1700–3400 m; different ecological preferences for each species; large, solitary, unarmed, pleonanthic, monoecious palms; stem tall stout or slender; leaves pinnate; inflorescence intrafoliar. Chromosome number not known.

In addition to the genera *Allalea* and *Syagrus,* the other genera that have been suggested to have relatively close relationships with the genus *Cocos* are *Jubaeopsis* and *Parajubaea*. Even a casual perusal of Figs. 3.3–3.5, and Table 3.1, shows that there is hardly any similarity in morphological and ecological characters even among these genera.

Figure 3.3 (A) *Syagrus vermicularisi.* (i) Habit, (ii) fruit (actual size 3–5 cm), (iii) nut, basal view, (iv) fruit in longitudinal section. (B) *Parajubaea cocoides.* (v) Habit, (vi) fruit (actual size 5–6 cm), (vii) nut, basal view, (viii) fruit in longitudinal section. *Reproduced from Dransfield, J., Uhl, N. W., Asmussen, C. B., Baker, W. J., Harley, M. M., Lewis, C. E. (2008).* Genera Palmarum. The evolution and classification of palms *(732 pp.). Kew, UK: Kew Publishing.*

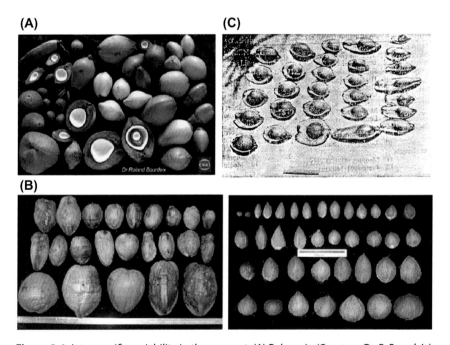

Figure 3.4 Intraspecific variability in the coconut. (A) Polynesia *(Courtesy: Dr. R. Bourdeix)*; (B) Laccadive Islands *(Courtesy: CPCRI, Kasargod)*; (C) Andaman and Nicobar Islands. *(Reprinted with permission from Balakrishnan, N. P., & Nair R. B. (1979). Wild populations of Areca and Cocos in Andaman & Nicobar Islands.* Indian Journal of Forestry, 2, 350–363).

(A)

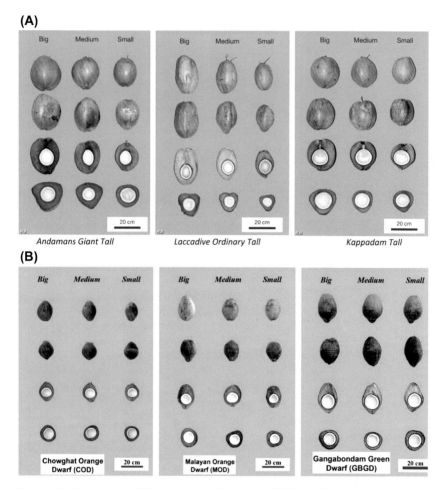

Figure 3.5 Range of variability in coconut landraces. (A) Variability in three tall landraces from India. (B) Variability in three dwarf landraces.

The genus is one of three genera that are stated to have the closest relationships with the genus *Cocos*. Noblick (2013) has recently reviewed the Subtribe Attaleinae. Its distribution is almost completely limited to South America. Presently, it includes c. 54 genera, 6 interspecific genera, and 1 intergeneric hybrid with *Butia*. In phylogenetic studies, the genus *Attalea* is often seen as a sister to *Lytocaryum* and *Cocos*. Incidentally, Noblick and Meerow (2015) have proposed the merger of the former with the latter. The *Syagrus* species is broadly distributed in two groups, Amazonian and Brazilian.

Table 3.1 Comparison of six subtribe Attaleinae genera including Cocos[a]

	Cocos	Attalaea	Jubaeopsis	Parajubaea	Syagrus
Species no. & chromosome no.	One sp. 2n = 32	c. 69 spp., 2n = 32	One sp, 2n = 160–200	3 spp. 2n = not known	31 spp. 2n = 32
Distribution	Worldwide, tropical, sub-tropical	Mexico, Caribbean, Bolivia, Peru	Southeastern S. Africa, rare	Ecuador, Colombia, Bolivia at high altitudes	Caribbean, S. America, mostly Brazil, in drier parts
Ecology	Strand plant; in humid weather, upto 900 m	Wide habitat range: Tropical rain forests to dry savannahs	Coastal reaches, gregarious, rocky banks of rivers	Humid ravines of cooler sandstone mountains at high altitudes, 2400–3400 m	Extremely variable, dry to semidry areas, acaulescent, conspicuous in dry vegetation types
Habit	Moderate, solitary, unarmed, pleonanthic	Solitary; small to massive	Clustering, branching at base with short erect stems, pleonanthic	Tall, stout or rather slender; solitary, unarmed, pleonanthic	Small to tall stem, solitary or clustered, pleonanthic
Fruit size, shape	Very large, 20–40 cm long; ellipsoidal to broadly ovoid, one seeded	9–10 cm long, 1–several seeded	c. 3.3 cm long; one seeded; globose with apical beak	5–6 cm long, oblong–ovoid, beaked; 1–3 seeded	2–5 cm long, small to relatively large; one-, rarely two-seeded
Mesocarp	Very thick, fibrous, dry	Fleshy or fibrous	Thick, fibrous; slightly fleshy	Thin, fibrous	Fleshy or dry edible
Endocarp	Thick, woody	Very thick; strong, smooth	Thick, bony, with 3 vertical grooves	Thick, very hard, with 3 prominent ridges	Thick, woody

Continued

Table 3.1 Comparison of six subtribe Attaleinae genera including Cocos[a]—cont'd

	Cocos	Attalaea	Jubaeopsis	Parajubaea	Syagrus
Endosperm	Homogeneous; large central cavity with partly filled water	Solid	Homogeneous with large central cavity, may be eaten	Homogeneous, hollow	Homogeneous, ruminate central cavity in some species
Leaf	Pinnate, neatly abscising	Pinnate	Pinnate in 5 vertical rows, marcescence or nearly abscising	Pinnate	Reduplicately pinnate
Stamens number	6	3–75	7–16	c. 15	6
Inflorescence	Monoecious; Solitary; interfoliar; auxiliary; branched to first order;	Monoecious; Staminate or pistillate or both on same inflorescence;	Monoecious; solitary; interfoliar; branching to first order, protandrous	Monoecious; interfoliar; erect, or pendulous; 1–2 order branching;	Monoecious; solitary; interfoliar; rarely spicate, usually branching to first order, protandrous (?)
Relationship	Moderately sister to *Parajubaea*, *Attalea*, and *Syagrus*	Sister to a clade of *Lytocaryum*; subclade of *Syagrus* with moderate support; monophyletic with high support	Sister to subtribe Attaleinae except *Beccariophoenix*	Moderately sister to *Cocos*	Polyphyletic

[a]Source: Various.

Tribe Cocoseae: Slender to very robust; acaulescent to erect to climbing; leaves pinnate or pinnately ribbed; inflorescence unisexual or bisexual, spicate to branched to one or rarely; two orders; usually within leaf sheath; gynoecium with three or rarely, more locules and ovules; fruit sometimes very large one to several seeded; endocarp with three or more pores; 18 genera in three subtribes, predominantly New World; characterized above all else by the presence of an endocarp with three or more clearly defined pores.

Subtribe Attaleinae: Acaulescent or erect, slender to very massive; flowers superficial. Includes all members of tribe Cocoseae except the spiny ones, *Elaeis* and *Barcella*; now, 11 genera, three in southeastern Africa and Madagascar, the rest in the New World.

Adapted from Dransfield, J., Uhl, N. W., Asmussen, C. B., Baker, W. J., Harley, M. M., Lewis, C. E. (2008). *Genera Palmarum. The evolution and classification of palms* (732 pp.). Kew, UK: Kew Publishing.

3. CHARACTER PRIMITIVENESS AND PROGRESSION IN THE PALM FAMILY

Moore and Uhl (1982), basing on their studies on the family Arecaceae, had proposed a pattern of primitiveness and progression in the evolution of plant organs and habits in the family. They are given below:

Habit: from sympodial to monopodial;

Size: from moderate to both large and small forms;

Stem: from unbranched to dichotomously branched;

Sclerenchyma: from short to elongate internodes;

Leaves: from undivided eophyll to palmate, costapalmate, or pinnately ribbed forms;

Inflorescence units: from moderately branched to spicate, or less frequently to more diffusely branched ones; from one unit in leaf axil to more than one per leaf axil, from among leaf axis to below the leaves; from pleonanthic to hepaxanthic forms;

Prophyll: from completely encircling to incompletely encircling the peduncle; from incompletely to completely sheathing bud;

Bracts: from conspicuous or small to absent at maturity; from peduncular bract, and from tubular and open at the apex to completely enclosing the inflorescence in the bud; and from ungrooved to deeply plicate;

Flower arrangement: from solitary, pedicellate, bracteate flowers to a sympodial cincinnus of two or more, or to a short monopodial axis of two to four or more flowers;

Bracteoles: from sheathing prophyllate to incompletely developed or absent;

Flowers: from bisexual to unisexual, then associated with polygamy or monoecism to dioecism;

Perianth: from trimery to dimery, or tetramery to decamery, or reduced and monochlamydous;

Sepals: from distinct and imbricate to connate or separated;

Petals: from distinct and imbricate to valuate or strongly imbricate or connate;

Filaments: from relatively slender and distinct to broad and thick, and often connate or adnate to the perianth or both;

Staminodes: from stamen-like with abortive anthers only to short teeth or to a cupule at the base of the ovary to absent;

Pollen: from monosulcate to trichotomonosulcate, to dicolpate to monocolpate, diporate or triporate;

Gynoecium: from apocarpous to syncarpus;

Carpules or locules: from three to two to one, or four to ten;

Ovules: from moderate to small or large, from anatropous to hemianatropous to campylotropous to orthotropous;

Pistillode: from only slightly modified from the gynoecium to vestigial or lacking, or rarely, to prominent;

Fruit: from fleshy to dry and fibrous;

Endocarp: from little differentiated or thin to thick and hard, and sometimes with a pore or operculum over the embryo;

Seed: from moderate to small or very large; from entire to dissected, bilobed, or perforate;

Endosperm: from homogeneous to invaginated or ruminate;

Germination: from remote tubular or ligular to adjacent–ligular;

Chromosome complement: from n = 18 to n = 17, 16, 14 or 13.

The authors had postulated that in the earliest beginning, five to seven major forms might have developed from a protopalm stock and that each of them appeared to have evolved into separate evolutionary lines.

4. INTRASPECIFIC VARIATION IN THE COCONUT

4.1 Introduction

Practically all of the authors who have studied this aspect have observed that the coconut crop displays much intraspecific variability. However, the range of variability that has been observed in the coconut does not appear any more than that found in any other cross-fertilized crop species.

Several authors had studied this aspect during the early decades of the 20th century, in keeping with the then prevalent practice in mostly temperate crops.

Watt (1889) studied the variability present in the coconut in his time in South Asia. He identified seven varieties. Cook (1901) reported 50 variant forms in Malaya (the present continental Malaysia). Burkill (1919) recorded 30 varieties in Malaya, and Jumelle (cf. Copeland, 1931), 25 variants in Java alone. Omar (1919) identified 12 landraces of coconut on Singapore Island, and Copeland (1931), 14 forms in the Philippines. There are numerous similar reports from several countries of South and Southeast Asia and a few from the Neotropics as well. Please refer to Gangolly, Satyabalan, and Pandalai (1957), Zizumbo-Villareal, Hernandez, and Harries (2005) and Zizumbo-Villareal, Fernandez-Barrera, Torres-Hernandez, and Colunga-Martin (2005), etc., for more information (Adkins, Foale, & Harries, 2002).

Bioversity International [BI, and its previous avatar, The International Plant Genetic Resources Institute (IPGRI), brought out the first standard description list for coconut way back in 1978. IPGRI has since been revising this list periodically. The present coconut descriptor gives also a core list of characters for quick characterization and evaluation of coconut germplasm. It had initiated steps to compile the full information on the coconut germplasm collected to date in 2008 (COGENT, 2010a). The work is being continued, but at a slow pace after withdrawing funding support by the BI. The data compiled in 2008 COGENT (2010a, 2010b) do not appear to have been collated. This publication provides some information on 116 accessories conserved in 27 countries.

The International Coconut Genetic Resources Network is an association of 38 countries promoted by the BI. The national coconut genetic resources database of different countries presently maintains 1416 accessions in 28 locations situated in 23 countries. It has also identified the 200 most important accessions in each region, with a view to maintaining them in multilocations (Batugal, Rao, & Oliver, 2005).

4.2 Features of Coconut Classification

4.2.1 Fewer Landraces

One characteristic of the coconut genetic resource variability, as compared to other crops, appears to be the presence of relatively fewer established landraces. The universal practice followed in the coconut crop, almost from the beginning, has been to name most common forms of a region after the locality from where each one has been collected; as for instance, Malayan Dwarf, Chavakad Dwarf (India) (Handover, 1919), Panama Tall, Fiji Tall (Marechal, 1928), etc. Some of the more well-known landraces are San Ramon (the Philippines), Gangabondam (India), Nyior Gadan (Malaysia), Coconino (the Philippines), Rath Thembili (Sri Lanka), Takame (Indonesia), Niu Kafa (West Polynesia), Kalapa Tebu (Java, Indonesia), Klapa Logo (Malaya), Niu Wa (New Caledonia), etc. A few more are listed in Section 4.3, later in this chapter.

In the coconut, what is termed in common literature as landrace/variety itself displays considerable variation within each in that almost every character, both quantitative and qualitative, usually displays a range of variation (Fig. 3.3). At the same time, such variant palms do give an overall impression of conforming to a broad morphological similarity. Hence, it may be better to designate them as morphotypes.

4.2.2 Talls and Dwarfs

A feature of the coconut crop is the presence of dwarfs. They are estimated to occur with 1–3% frequency in natural populations. Generally, the dwarfs show features of partially miniaturized forms of normal tall palms, say, by 50–60% in all respects, that include tree height, time of first flowering, longevity, organ size including fruit size, etc. Swaminathan and Nambiar (1961) had suggested that dwarfs arose by inbreeding of talls. Recently, Gunn, Baudouin, and Olsen (2011), too, had supported this. However, this has since been found to be incorrect (Raveendranath & Ninan, 1974). The dwarfs appear to be naturally occurring short-statured mutants.

Perera et al. (2016) has just come out with a finding that "dwarf coconut originated from a typical [sic] domestication event in southeast Asia." They concluded this using SSR allele frequency, differences between tall and dwarf accession ethnobotanical information.

The authors calculated allelic frequencies using ARLEQUIN program in 51 tall and 43 dwarf varieties representing a global collection. They analyzed also 246 belonging to 28 dwarf accessions with microsatellite markers. The data were analyzed by computing Euclidean distance for each locus and summing over all loci, which ensures that the completely homozygous allele is equal to the number of allele substitutions. The dwarfs had fewer alleles and one allele was not observed in talls. In many instances, both dwarfs and talls had the same alleles, but it was not so with four loci. The "most frequent allele" in the dwarfs was rare in the talls. In a similar manner, the most frequent allele in the talls was not present in the dwarfs. The dwarfs shared 43 alleles with the Indo-Atlantic group and 22 alleles with the Pacific group. In the dendrogram, all the dwarfs were seen embedded in a Southeast Asian and Pacific cluster. Further, most were grouped in a single subcluster. The authors took this to mean that "it confirms that the Pacific group is the origin of dwarf varieties."

The authors have not given any ethnobotanical and specific geographical information in support of their conclusions, as mentioned in the Summary.

4.2.3 Micros

A third feature is the occurrence of "micros"—palms producing nuts that are only about half or even less the size of normal coconut fruits and nuts. Some examples are: Laccadive Micro, Andamans Micro, Coconino (Philippines), Laccadive Mini–Micro, Ayiramkachi (literally meaning "thousand (fruit) setter," a landrace from Kerala, India), etc. There are more such forms, especially in the Indian and western Pacific Ocean islands. This form may have an evolutionary relevance, as we shall see later.

4.3 History of Variety Classification

4.3.1 Classical Classification

Taxonomic classification. Narayana and John (1949) were the first to attempt the classification of the cultivated coconut using classical taxonomy. They first classified it into five taxonomic varieties, var. spicata, var. typica, var. androgena, var. japonica, and var. nana. Then they divided the variety "typica" into nine forms: pusila (Laccadive small), ramona, kappadom, gigantea (Andaman Giant), siamea, nova-guineana, malayensis, cochin-chinensis, and laccadive. Their characteristics were given as under:

spicata: inflorescence generally unbranched, rarely one to two branched

typica: tall, late bearing

androgena: only male flowers

javanica: vigorous, bearing in 4 years

nana: delicate, bearing in 3 years

Variety "typica" was further classified into "forms."

pusilla: c. 1000 cc volume, 100 or more fruits/bunch, e.g., Laccadive small;

gigantea: fruits 7000 cc in volume; copra content low and of poor quality, e.g., Andaman giant;

ramona: fruits c. 6000 cc or less in volume, superior and high-quality copra, yield 100 nuts/year;

kappadan: (sic, Kappadom): fruits c. 6000 cc or less; yield low, 35 nuts/year;

siamea: fruits c. 4000 cc or less in volume. trunk robust 90–100 cm in girth, high (74%) oil content;

nova-guineana: fruits c. 4000 cc or less in in volume, robust trunk (90–100 cm) and low oil content (66–69%); high female flower production (744 nos.);

malayensis: nuts c. 4000 cc or less in volume, robust trunk (90–100 cm girth), low oil content (66–69%); female flowers low (220 nos.);

cochin-chinensis: trunk medium girth (73–83 cm); copra thin (8.2 mm); low oil content (66%);

laccadive: trunk medium girth (73–83 mm); copra thick (12–13 mm); high oil content (72%).

Incidentally, Jacob (1941) had first described the botanical variety spicata; in addition, the paper Narayana and John (1949) was published again under the authorship, John and Narayana (1949).

This work was followed by several attempts in some countries—notably, India, Sri Lanka, the Philippines, and Mexico—to characterize/catalog the variability present in their respective countries (Gangolly et al., 1957; John

& Satyabalan, 1955; Liyanage, 1958; Menon & Pandalai, 1958; Parham, 1960; Whitehead, 1966; Zizumbo-Villareal, Fernandez-Barrera, et al., 2005; Zizumbo-Villareal, Hernandez, et al., 2005).

It appears to have been realized soon that attempts to categorize the extant variability in different regions/countries was impractical/impossible, especially into taxonomic entities. This was because, although there was no difficulty in categorizing the palms as tall and dwarf, the normal tall forms show a complete range of variation for virtually all the characters such as nut size, color, shape, water content, various leaf characters, and so on.

The collective experience in other crops indicates that subspecific classification of cultivated species into taxonomic categories as subspecies, variety, etc. is impractical, and of only limited relevance. The best—and possibly the most studied—instance has been the attempt to classify Asian cultivated rice, *O. sativa,* by Kato, Kosaka, and Hara (1928). Kato et al. had initially recognized two subspecies, indica and japonica. Despite both very extensive and intensive investigations done by numerous workers over several decades for delimiting their taxa boundaries, all the attempts failed to satisfy the taxonomic criteria for determining them as subspecies (cf. Nayar, 2014; Oka, 1988, for details). Although in cultivated rice, it is possible to identify typical indica and japonica forms, the majority of landraces possess attributes of both indica and japonica forms in varying proportions. In rice, japonica and indica forms are best considered ecotypes. This is because, very broadly, typical indica rice occur in the warmer moist regions of tropical South Asia, whereas typical japonica rice occur in the cool termperate regions as in Japan.

4.3.2 Ecotype and Morphotype
In the coconut, it is inappropriate to use the epithet ecotype, because as a lowland tropical plant species, the coconut palm does not appear to have undergone comparable ecological differentiation as rice has. Hence, the suggestion here is to designate them as morphotypes in the coconut.

At the same time, some morphotypes occur in different countries either autonomously or after selection for certain attributes for extended periods by humans. And, they show broad similarities. They include forms having (1) large fruits: e.g., San Ramon (Phillippines), Kappadom (not Kappadan, as given in some literature; Kerala, India), Andamans Giant (Andamans, India), Ramona (Philippines), Markham (New Guinea); (2) dwarf forms: e.g., Chavakkad Dwarfs, three forms (Kerala, India), Malayan Dwarfs, three colors (Malaysia), Fiji Dwarf (Fiji), and so on; (3) forms with very small fruits: e.g., Lakshadweep Micro (Lakshadweep, India), Coconino (Philippines), etc. More examples are given in Gangolly et al. (1957). It is proposed that such forms alone may be identified as landraces in the coconut.

4.3.3 Molecular Biology Characterization

The few comparative genetic or molecular genetic studies carried out to understand the relationships among the morphotypes have hardly been able to provide any insight into the mode or direction of differentiation of the coconut (Devakumar et al., 2006; for instance). Most of the over 20 studies carried out in the coconut using molecular biology techniques have been done more in the nature of testing the suitability of various techniques, and establishing the priority of the authors in the use of a particular technique. The main weakness of these studies has been that the material used in them had been chosen arbitrarily, and not comprehensively and selectively except in a few, e.g., Ashburner, Thompson, Halloran, and Foale (1997, all the Pacific variability; see the next paragraph).

The method used by Whitehead (1966, 1968) using fruit component analysis appears to have been useful only to a limited extent. Rathnambal et al. (1995) data show that all the characters, including the nut characters, show much variability. This study is incidentally the first comprehensive one carried out to prepare a catalog of coconut germplasm. It is the most comprehensive study done to date for characterizing the coconut germ-plasm. The authors had employed the minimum descriptor list of BI. It involved characterizing 13 vegetative, 21 reproductive, 20 nut, and 2 bio-chemical attributes. This work was done in 72 accessions involving a world collection. The data were published in two parts: 48 nos. in print form and 24 nos. in interactive CD format. Part of these data has been used to prepare Table 3.2.

Ashburner et al. (1997) studied Pacific coconut variability using a representative collection of the entire Pacific coconuts in an attempt to determine their interrelations. They used 29 accessions and they grouped them into four sets. The authors observed large diversity in fruit mor-phology. It ranged from populations exhibiting wild type characters in the Central Pacific collections to those having domesticated characters in those originating in Rennel Island (11°40′ S, 160° 10′ E), Marquesa Island (9°30′ S, 140° 0′ W), Sikaisna atoll, Solomon Islands (8°38′ S, 162° 71′ E), and PNG. Several populations displayed intermediate characters. Cluster analysis (canonical variate analysis) "arbitrarily divided" the

Table 3.2 Fruit and nut character analysis of a world coconut collection (47 nos.)

Sl. No.	Region/variety name	Fruit set (%)	Bunch set (%)	Fruit length (cm)	Fruit L–B ratio	Tender nut water (mL)	Fruit weight (g)	Husk weight/fruit weight (%)	Kernel weight (g)	Shell weight (g)	Nut weight (g)
I	**Micronesia**										
1	Guam I	31	75	33	1.9	290	1214	38.3	405	158	747
2	Guam II	25	75	32	1.8	215	1369	37.8	462	164	850
II	**Melanesia**										
3	Solomon islands	41	67	28	1.6	356	879	35.9	322	133	566
4	Nufella	26	NA	23	1.7	170	613	38.2	214	100	378
5	Nugili	20	97	30	1.9	270	580	38.3	202	94	360
6	Nuqeawen	26	NA	29	1.9	260	636	37.0	223	115	400
7	Nuwehung	20	91	29	1.9	250	667	41.6	215	100	388
III	**Polynesia**										
8	Fiji Longtongween	22	67	31	1.6	390	981	0.63	298	183	618
9	Fiji Tall	25	86	30	1.3	330	905	44.2	287	127	505
IV	**Papua New Guinea**										
10	NG Tall	6	60	26	2.1	358	1050	39.3	302	141	635
V	**Philippines**										
11	P. Laguna	43	69	34	1.9	260	1059	35.1	356	186	687
12	P. Lono	22	77	31	1.6	340	1509	47.6	338	214	795
13	P. Ordinary	28	68	28	1.8	452	1031	35.7	357	117	665
14	San Ramon	22	82	39	2.0	890	1998	30.5	600	285	1389
VI	**Southeast Asia**										
15	Borneo Tall	27	73	31	1.5	625	1450	31.6	444	172	987
16	Fed. Malay states	22	55	31	1.8	600	1130	0.67	358	141	758
17	Java Tall	29	70	33	1.6	495	1155	34.1	388	152	761

No.	Cultivar									
18	Kongthienyong	23	65	31	1.6	390	1381	36.3	484	187	881
19	Malayan G dwarf	34	55	22	1.3	290	724	0.55	283	119	396
20	Malayan O dwarf	25	50	29	1.8	303	865	35.4	290	123	558
22	St Settlements green	25	50	31	1.7	330	1130	32.8	358	167	758
23	Andamans giant (A & N)	35	45	33	2.2	290	1201	42.3	331	157	693
24	Andamans Ordinary (A &N)	14	47	31	1.9	274	1182	47.6	300	159	620
25	Ranguchan Andamans	25	45	33	2.5	345	1204	37.3	367	221	757
VII	Indian Ocean islands										
27	Ceylon tall (S Lanka)	24	80	30	1.8	275	1092	40.9	328	188	648
28	Gonthembili (S Lanka)	26	84	31	1.5	274	957	45.2	280	163	524
29	King coconut (S Lanka)	31	79	25	1.4	358	447	0.64	0.37	0.37	0.18
30	Seychelles tall	15	74	33	1.8	175	700	58.5	179	100	292
31	Zanzibar tall	47	97	28	1.6	348	861	33.9	314	129	569
32	Laccadive micro (A & N)	40	42	20	1.8	75	562	49.2	156	62	282
33	Laccadive Ordinary tall	40	47	34	1.6	285	720	39.1	274	118	438

Continued

Table 3.2 Fruit and nut character analysis of a world coconut collection (47 nos.)—cont'd

Sl. No.	Region/variety name	Fruit set (%)	Bunch set (%)	Fruit length (cm)	Fruit L–B ratio	Tender nut water (mL)	Fruit weight (g)	Husk weight/fruit weight (%)	Kernel weight (g)	Shell weight (g)	Nut weight (g)
IX	**India, west coast**										
34	West coast tall	38	79	26	1.2	241	1196	52.3	283	145	566
35	Benaulim (Goa)	36	56	24	1.7	250	799	36.0	253	128	512
36	Calangute (Goa)	22	46	33	1.6	280	940	47.1	316	130	496
37	Nadora tall (Goa)	33	57	30	2.2	290	862	37.1	300	132	536
38	Chavakad green dwarf	38	32	25	1.4	190	453	58.5	121	65	190
39	Chavakad orange dwarf	30	36	22	1.7	351	634	40.7	194	93	379
40	Gujarat Zanzibar	17	62	33	1.8	270	1250	34.9	400	188	814
41	Kappadam tall	24	80	34	1.8	425	1200	30.1	505	160	816
X	**India, rest of**										
42	Gangabondam, Andhra Pradesh (AP)	30	40	29	1.9	267	804	36.7	266	135	511
43	Kenthali dwarf (Karnataka)	26	24	27	1.8	206	456	29.8	193	102	323
44	Rangoon Kobbari (AP)	26	59	33	1.6	404	1502	40.7	467	173	890
45	Sakhigopal (Orissa)	35	47	26	1.7	237	881	39.2	284	130	537
XII	**West Indies**										
46	Jamaica Sanblas	33	98	28	1.5	263	920	31.6	304	142	630
47	Jamaica tall	11	78	31	1.7	362	1427	43.8	400	193	802
48	St Vincent (Trinidad)	15	77	30	1.5	306	1128	55.4	245	185	503

Basic data from Rathnambal, M.J., Nair, M.K., Muralidharan, K., Kumaran, P.M., Rao, E.V.V.B., Pillai, R.V. (1995). *Coconut descriptors – I* (pp. 197). Kasaragod, Kerala, India: Central Plantation Crops Research Institute.

continuum into four discrete groups, which were consistent with geo-graphical affinities, viz., Melanesia, western Polynesia, eastern Polynesia, and PNG. They displayed clonal variations with populations having small fruits and low husk content in the west to large fruits and more husk content toward the east. Intriguingly, the wild and domesticated populations were found in disjunct pockets throughout the region, and they did not form part of any cline. Most populations, at the same time, showed within each of them a whole range of characters from wild to domesticated forms.

4.3.4 Niu Kafa and Niu Vai

Based on the analyses carried out by certain early workers (Copeland, 1931; Dwyer, 1938; Narayana & John, 1949; Parham, 1960; Whitehead, 1966). Harries (1978) introduced these two terms for the classification of the coconut. Whitehead (1966), on the basis of the coconut collections made in the Pacific islands, had stated that some of the "more isolated" islands—such as Rennell (Solomon islands), Rotuma (Fiji islands), and Wallis (Wallis and Futuna Islands, 13° 18′ S, 176° 10′ W)— had palms producing nature nuts with 600–700 cc water and high kernel content (quantity not given). They were locally known as Niu Vai.

Such fruits (nuts) also occurred on other islands such as Samoa and Tonga. Whitehead then proposed that this form, having a substantial quan-tity of water (600–700 mL), might have been selected by early Polynesian navigators for drinking water. According to Harries (1978), similar forms occur in several other countries also: e.g., Thifow in Micronesia, Bali (Indonesia), (Kappadam/not Kappaden; Narayana & John, 1949), Rangoon Kobbari (Myanmar; Gangolly et al., 1957), Kamandala (Sri Lanka; Liyanage, 1958), Lupisan and Sam Ramon (Philippines; Copeland, 1931), and Markham (New Guinea; Dwyer, 1938). At the other end, Harries (1978) observed that niu kafa variety, the primitive form, consisted of long, angu-lar nuts with thick husk, and showed slow germination. He stated that "introgression" (natural crossing) takes place between the two extreme types in nature, wherever they occur together as in Western Samoa. Niu kafa, he proposed, "was naturally evolved," and niu vai was "selected under cultivation."

Harries (1981b, 1982) refined this concept in two subsequent papers after he attempted to apply them on already published accounts of vari-eties representing different regions of the world and the Indian collec-tions, respectively. In his 1981 paper, he observed that niu kafa forms evolved by natural selection on uninhabited coral islands and newly emerged volcanic islands. Niu kafa was characterized by long angular thick-husked fruits and slow seed germination, with these factors favoring "survival under natural conditions in the absence of man." It has only a "very restricted habitat in the natural state. They fringe the shores just above the water mark, where they are not shaded by forest trees, nor choked by undergrowth."

The niu vai forms were used "to provide sweet uncontaminated drink just this purpose by the first human beings… Selection increased the volume of the liquid endosperm in the immature fruit at the expense of the other fruit components, principally the husk." These forms were "marked also by a more spherical appearance, early germination, and more vigorous growth …," all of which were the result of selection under domestication. He opined also that West African Tall and Mozambique Tall were similar, Tahiti Tall was superior, and Malayan Tall was intermediate between the two. The author appeared to face difficulties in interpreting the extant variation in the characters of the geographically diverse accessions that he had analyzed.

Harries (1982) reexamined some of the published literature on Indian varieties. He then appeared to have firmed up by then his views on niu kafa and niu vai forms. He defined niu kafa type as "evolved by natural selection without human intervention" and niu vai type as "domesticated by primitive man for the sweet uncontaminated water in the young fruit". Both were independently taken into cultivation. The former possesses thick husk and niu vai has thin husk and heavier nut size. Harries then defined the nut characteristics of the two forms as follows:

Niu kafa: 15–25% water, 27–36% shell, 46–52% meat

Niu vai: >35% water, >24% shell, <42% meat.

The speed of germination was similar for both forms. Harries added further that norms for niu kafa and niu vai types would be different in different countries, apparently when he found it difficult to group all the populations into these two forms.

It is apparent that Harries has been altering his concepts on the two forms. A perusal of the literature indicates that there is general agreement that the primitive coconut forms have long angular fruits and nuts, their husk is thicker, and the whole fruit and nut sizes are smaller. Generally, scholars are willing to accept the designation of nui kafa for primitive forms, but this cannot be stated for his concept of niu vai, or the idea behind it. A perusal of Table 3.2 will show that the stated indicators do not hold well in them (cf. also Section 4.5 in following section).

4.4 Variety Classification

In the early years, various authors had reported variations in the size of the coconut fruit and its components. We refer to them here as morphotypes. They have been summarized by some authors (Gangolly et al., 1957; John & Narayana, 1949; Patel, 1937, etc.). A few of the more distinctive forms are listed here for illustrative purposes.

A. *Cocos nana* Grif.: from South Asia; a very small dwarf variety going by the name Maladive coconut (Trimen, 1898). It appears similar to what is now referred to in literature as Maldives Micro Tall.

B. Tembili from South Asia: mentioned by Watt (1889), from India; Trimen (1898) from Sri Lanka: possesses pink endosperm; more commonly known as king coconut; a dwarf form.

C. Klapa Wangi: from Singapore (Omar, 1919); has fragrant endosperm; mentioned also by Burkill (1935).

D. Sam Ramon: from the Philippines (Copeland, 1931); high yielding with large nuts; similar types reported from India, Sri Lanka, Malaysia, and Polynesia.

E. Coconino: from the Philippines (Copeland, 1931); also known as baby coconut; very dwarf in nature, possibly similar to micro and nana forms.

F. Pugai: from the Philippines (Copeland, 1931); very dwarf palm with short juvenile period.

G. Makapuno: from the Philippines (Copeland, 1931); instead of the usual firm endosperm, a light curd-like endosperm fills the cavity; considered a delicacy; *Thairu thenga* (literally, curd coconut) of Malabar Coast (India) and Dikiripol (Sri Lanka) appear similar, and possibly independently evolved.

H. Spicata: described originally as a taxonomic variety, *C. nucifera* var. spicata (Jacob, 1941); inflorescence unbranched with no spikelets present, bears c. 100 female flowers and 50 male flowers in an inflorescence; produces 10–20 mature fruits/inflorescence.

I. Male coconut: The inflorescence bears only male flowers and no female flowers, with c. 5000 male flowers per inflorescence (John & Narayana, 1942).

J. Two of the "freaks and abnormalities" listed by John and Narayana (1949) are viviparous and suckering forms. The others are apparently teratological abnormalities;

K. Mangipod: intermediate between Coconino and Pugai; has very short juvenile phase (Copeland, 1931).
L. Niu Leka: the dwarf coconuts of Fiji; a prolific bearer.
M. Nudam: from Fiji; husk "exceptionally" thin; hardy in nature.
N. King coconut: also known as Rath Thembili; from Sri Lanka; juice very sweet; esteemed for culinary purposes.
O. Taban: from the Philippines: similar forms are known as Cayamis in northern Mindanao, and Kalapa Tebu in Java; husk sweet when tender, which can be chewed like sugarcane.
P. Kaithathali: a tall form from Malabar Coast, India; mesocarp soft and fleshy, sometimes eaten; said to be a good antidote for seasickness (Patel, 1938).
Q. Navasi: from Sri Lanka; mesocarp is sweet.
R. Coco raisin: from Seychelles; nuts small and spherical, "resembling grapes".
S. Klapa Dadeh: from continental Malaysia (Omar, 1919); the water within the endocarp cavity contains endosperm in granular form; possibly similar to or same as Makapuno.
T. Klapa Logo: from continental Malaysia (Omar, 1919); nuts are edible at the tender-nut stage.
U. Klapa Angi: from continental Malaysia (Omar, 1919); endosperm possesses pleasant smell; used in preparing medicines.
V. Nu Wa or Cocos suere: from New Caledonia; mesocarp sweet and edible like sugarcane.

An interesting observation is the occurrence of similar morphotypes in various major growing countries: an instance of parallel variation. A few instances are given as follows:

A. Those yielding big nuts: San Ramon (Philippines), Kappadom (Kerala, India); similar forms also reported from Malaysia, Papua New Guinea, etc.;
B. Those having curd-like endosperm: Makapuno (Philippines), Thairu Thengai (India), Klapa Dadeh (Malaysia);
C. Very dwarf forms with short juvenile phase: Coconino, Pugai (both Philippines); Laccadives Mini–Micro (India);
D. Those having sweet and edible mesocarp while young: Taban (Philippines), Kaithathali (Kerala, India), Klapa Logo (Malaysia), Nu Wa from New Caledonia.

This is not a comprehensive list. There are more like these.

4.5 Cataloguing the Variability

The Central Plantation Crops Research Institute (CPCRI), Kasaragod, India, maintains a coconut germplasm collection of 438 accessions. It includes 306 collections from outside India. The institute has an ongoing program to catalog the collection. So far, it has published two volumes of the catalog covering 72 accessions (Rathnambal et al., 1995); using BI's minimal list of descriptors consisting of 56 characters—consisting of 13 vegetative, 21 reproductive, 20 nut, and 2 biochemical characteristics.

Excerpts from the first two volumes representing a near-global collection relating to selected fruit characters are given here (Table 3.2) for 47 accessions along with some analyses.

All the accessions are cultivated forms. The details of the procedure followed in collecting the primary data are given in the publication. A notable feature is the very wide variation noticed for all the characters in all the accessions studied here. Thirdly, almost all of them show features associated with various degrees of ennoblement.

The previous observations indicate that the objectives of ennoblement have been different in different regions of the world. Some characters have not shown any indication of ennoblement, when we compare the situation with some of the oldest cultivated plants. For instance, the highest fruit set obtained in the coconut was only 47% (in Zanzibar Tall). In the cereals, the comparable values would be 95–100%, and only rarely lower. Even a slight reduction of about 5% usually happens only because of the influence of some environmental factors such as water deficiency, pest or disease incidence, nutrient deficiency, and so on.

A. **The fruit set** ranged from 6% to 43%. The 6% setting reported in New Guinea Tall may be possibly an aberration/error. The next-lowest fruit set was obtained in Seychelles Tall (22%). Most of the setting were in the range of 22–35%.

B. **The bunch set** (the number of bunches produced in an year vis-à-vis the number of inflorescences produced in that year) ranged 24–97%.

C. **The fruit weight** ranged from 370 g (Malayan Yellow Dwarf) to 1998 g (San Ramon, the Philippines).

D. **The volume of tender nut water** ranged from 75 mL (Laccadive Micro, Laccadives, India) to 890 mL (San Ramon). The majority of the varieties contained 250–350 mL water. The mature nut water content was usually in the range of one-third to one-fourth of their respective tender-nut water content.

E. **The fruit length** varied from 20 cm (Laccadive Micro, India) to 39 cm (San Ramon). Majority of the accessions were in the range of 28–33 cm.

F. **The length–breadth ratio of fruits** ranged from 1.3 (Fiji Tall) to 2.5 (Andamans Ranguchan, India). Majority of the accessions were in the range 1.5–1.9.

G. **The husk weight** ranged from 29.8% (in Kenthali Dwarf, Karnataka, India) to 58.5% (in Chavakad Green Dwarf, Kerala, India). The husk weight of most accessions was in the range, 30–38%.

H. **The kernel weight** ranged from 121 g (in Chavakkad Green Dwarf, Kerala, India) to 600 g (San Ramon, the Philippines). Majority of the accessions were in the range of 250–350 g kernel/fruit. The shell weight mostly ranged from one-third to one-half of the kernel weight. Broadly, the kernel weight came to about one-third the weight of the fresh fruit.

4.6 General Observations

Broadly, the fruit size of the collections from Melanesia is smaller than that of the rest; and that from the Philippines and PNG are bigger. The fruits of dwarf forms are small to medium, except for the "green" forms, which are generally smaller than the other dwarf forms, "orange" and "red." Chavakkad Green Dwarf is smaller than all the other dwarf green forms.

The very limited data presented here do not indicate any clear-cut directions in the evolution of ennobled forms of coconut. Analyses of more extant morphotypes representing more regions and more samples/accessions will be required for understanding this. There are some indications that selection for bigger nut size and higher endosperm content has happened in the Philippines, for high nut number in the Laccadives, and for multipurpose forms in India, particularly on the Malabar–Konkan Coast. The nut and fruit sizes of accessions from West Indies and West Africa are also generally bigger and bolder, but we are unsure if this is reflective of the initially introduced forms, or a consequence of intensive selection carried out in the region, as the coconut culture in these regions is only 500 years old—even though, 20–25 generations are enough time for selection for desired characters under domestication. This region did not have any primitive or wild (or niu kafa) forms with which the introduced forms could have cross-pollinated, as it might have been happening in much of South and Southeast Asia. Further studies are needed to confirm these preliminary indications. One aspect that may be unique in the coconut is that only the fruit characters show differences among the (so-called) coconut varieties, and not apparently for other plant characters, except for the gross morphological differences between the tall and dwarf forms.

CHAPTER 4

Paleobotany and Archeobotany

1. INTRODUCTION

Paleobotany is the study of fossil plants. A fossil plant is the remains or traces of a once living plant (Allaby, 2006). Fossil plants are generally found buried below ground.

Paleobotanical information is used to unravel the evolutionary history of plant taxa, in both time and space. It is employed also as a benchmark in phylogenetic studies for estimating differentiation times of different levels of taxa. The classical example of the successful application of paleobotany in this respect has been the estimation by Daghlian (1981) of the time of differentiation of grasses as 50–60 Mya from the other angiosperms. He estimated this age by observing the changes in the dentures of ancient grazing animals in the South American grasslands. Various later studies employing advanced methodologies have largely confirmed this date (Doebley, Gaut, & Smith, 2006; Gaut, 2002; for instance).

Another application of paleobiology is in monitoring major climate changes. The chronological expansion and retraction of the distribution of fossil palm records at higher altitudes—particularly, during the cooling events of the Early Miocene (22.5–20 Mya) based on fossil evidence—is often taken as an indicator of drastic global changes having taken place in the past.

The paleobotany of the Arecaceae family has been reviewed in the past from time to time (Dransfield et al., 2008; Harley, 2006).

The Arecaceae family has produced the most extensive fossil records among all the monocotyledons (Harley, 2006). They are geographically widespread and comprise a wide range of organs and periods. The most frequent recovered organs are the leaves, stems, and pollens. However, most of the fossils lack sufficient diagnostic details to allow even a reasonable association with extant palm taxa, even to the subfamilial level (Dransfield et al., 2008; Harley, 2006).

Based on fossil evidence, Africa and India appear to have once possessed much richer palm floras than at present. Now, both regions are considered palm-poor (Harley, 2006). She observed also that the fossil records of the

The Coconut
ISBN 978-0-12-809778-6
http://dx.doi.org/10.1016/B978-0-12-809778-6.00004-8

51

Family were "extraordinarily rich and diverse from Palaeocene (66–57 Mya) to the cooling events of Miocene (22.5–7.0 Mya)."

The earliest unequivocal fossil palm material dates from early to mid-Late Cretaceous (141–106 Mya). It consisted of costapalmate leaves of *Sabalites caroliensis* from middle-eastern USA (Berry, 1914). Pinnate leaves—as present in the coconuts—were first described from northern Montana, USA, during the mid-Upper Cretaceous (100–70 Mya) (Crabtree, 1987).

Fossil stems and leaves are often difficult to identify below the family, or at the most, subfamily level. It is somewhat better with pollens. There are also some reports of *Cocos*-like pollen. Most of the fossil reports of *Cocos*-like fruits and nuts have come from just two regions in the world—New Zealand and west-central India.

2. FOSSILS FROM NEW ZEALAND

The earliest fossil report of *Cocos*-like fruits was from the Miocene age (23.0–5.3 Mya) (Berry, 1926). Berry was sent a sample collected from Mangonui, northern Auckland, New Zealand (35°1′S, 173°32′E) (Fig. 4.1). The nut was relatively small (3.5 × 1.3–2.5 cm) and spheroidal in shape.

Phylum:	PLANTAE
Class:	Spermatopsida
Order:	Palmales
Family:	Palmae
Subfamily:	Cocoideae
Genus:	*Cocos*
Species:	*Cocos zeylandica*
Locality:	Cooper's Beach, Mangonui Doubtless Bay, Northland, N. Z.
Age:	Waiauan Stage - Middle Miocene
Preservation:	Carbonaceous.
Collected:	Michael K. Eagle – 15.12.2000
Common name:	coconut

Figure 4.1 Fossils of *Cocos zeylandica* recovered from Mangonui, near Auckland (actual size 3.5 x 1.3–2.5 cm). *Reproduced from University of Waikato, New Zealand, with permission.*

They were invariably flattened in preserved state. The entire fruits were only slightly bigger than their nuts. The nuts were impregnated with pyrite (an iron ore, FeS_2). They contained three subequal foramine (basal pores), about 7 mm apart and c. 3 mm in diameter. The "husk is of the typical loosely fibrous *Cocos*-type, and in preserved state, 5 mm thick. The palm fruits show (showed) no features by which it[sic] can (could) be different from the existing genus *Cocos*" (Berry, 1926). The locality where the fruits and seeds had been collected was the Kaikorai beds in Oamaru district containing brown coal. It appeared to be of Miocene age (23.0–5.3 Mya), and representing the pre-Pliocene–Tertiary of New Zealand (please see following paragraphs). Berry named the material, *Cocos zeylandica*.

Couper (1952) observed that fossil coconuts from Coopers Beach, Mangonui (Auckland, New Zealand) were very well preserved and that they were "commonly washed up on the beach." Ballance, Gregory, and Gibson (1981) reported the recovery of the coconut (*Cocos zeylandica*) from the Miocene of Northern Island, New Zealand. They recovered the material from the proximal turbidites (a sediment or rock deposited by a turbidity current). The authors felt that the turbidites might have been formed by a surge of sediments and debris carried offshore by tsunamis in the geological past. They first located the nut-bearing beds in flysch (a sedimentary deposit consisting of thin beds of shale alternating with coarser strata, such as sandstone). A second nut was recovered in a loose block less than 1 km away. The flysch was more than 200-m thick and 10 km^2 in area. The authors estimated the age of beds to mid-Miocene (15–12 Mya) or younger, basing the age on benthic and plankton fauna found along with them. They, too, identified the nut as *Cocos zeylandica*.

Hayward, Moore, and Gibson (1960) stated that till then small *Cocos*-like nuts had been recovered from four locations in New Zealand. They were from (1) Waikiekie quarry, west Bryndervyns, North Island. This quarry of sandy or muddy limestone had been dated to Oligocene (36–23 Mya). The nuts had been recovered in 1982. The others consisted of (2) scattered occurrences perioded to Eocene (57–36 Mya), Oligocene (36–23 Mya), and Miocene (23–5 Mya) sedimentary strata from Northland; (3) Hawkes Bay, North Otago; and (4) South Canterbury (Ballance et al., 1981; Berry, 1926; Campbell et al., 1991). In all the cases, the nuts were generally 5-cm long with three distinct eyes at one end.

Hayward proposed that the all the above nuts belonged to the temperate genus, *Parajubaea*. The species of this genus occurs in northern South

America—at high altitudes (2500–3000 m above msl)—in Ecuador and south Colombia (Endt & Hayward, 1997). Incidentally, the senior author is a plant collector and nurseryman. They observed further that the fossil nuts showed most resemblance to *P. coccoides*, which is one of three species of this genus. *Parajubaea* has thick hard shells with one to two seeds in a fruit, and rarely three. The fruits of all the *Parajubaea* species are 3–6 cm long.

Another finding was of Campbell et al. (1991). According to them, one Park had reported in 1886 that he had observed numerous palm nuts along the Oamaru–Livingstone rail line, near the Livingstone tunnel (c. 100 km north of Dunedin; 45°05′S 170°59′E), which was then under construction. The nuts were 35–45 mm in diameter. They occurred in glauconite horizons dated to Middle Eocene (c. 50 Mya) or Middle Miocene (15–12 Mya).

Black (1996) examined a riparian debris upstream on Napier–Taupo river (Central North Island, New Zealand), along State Highway 5. He obtained two samples of *Cocos zeylandica* on the surface. They consisted of upper and lower halves of a nut. Another part husk was found also along with fossils of gastropods and mollusks, which implied a shallow marine environment. They belonged to Miocene period (23–5 Mya).

We can see that whereas most of the materials belonged to the Miocene epoch, there is also one report each of nuts belonging to Oligocene and Eocene epochs.

3. FOSSILS FROM INDIA

A second region that is rich in fossils of *Cocos*-like fruits, nuts, and stems is west-central India, and the region lying to its north [covering the present-day Madhya Pradesh (MP), Rajasthan, and Maharashtra]. This region is noted for the presence of Deccan traps (Fig. 4.2).

The Deccan traps is a large igneous province located on the Deccan plateau in west-central India (17–24°N) (Wikipedia, 2015). It consists of multiple layers of solidified basalt. They are more than 2000 m thick and presently cover 0.5 M km² area. It began to form 66 Mya at the end of the Cretaceous (144–66 Mya) through a series of volcanic eruptions that lasted about 30,000 years. This took place from the end-Mesozoic into early Tertiary era. The Trap had originally covered 1.5 M km² area, but now, its extent has been reduced to more than half by erosion and plate tectonics. In-between the layers of solidified basalt, there are deposits of

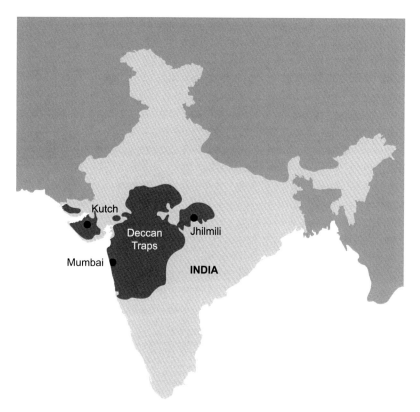

Figure 4.2 India showing Deccan Traps. *Reproduced with permission of Princeton University, USA.*

variable thicknesses consisting of soil with fossil flora and fauna remains. These deposited layers are termed intertrappean formation. The Deccan traps is slated to be the most extensive geological formation of this kind in the world (Wikipedia, 2015). There is, however, some disagreement about the period of formation of the Deccan traps. Morley (2000) has proposed that they could extend from Upper Cretaceous to mid–Oligocene (c. 145–38.1 Mya).

Sahni (1946) recovered a silicified stem of a *Cocos*-like palm, *Palmoxylon sundaram* from the Deccan intertrappean series. Then, Mahabale (1958) reported presence of another taxon, *Palmoxylon insignae*. Kaul (1951) obtained the fruit of a palm (which he named *Cocos sahnii*) from a fuller's earth mine in Kapurdi (25°54′N, 70°22′E), 19 km north of Barmer, near Jodhpur, Rajasthan. The nut showed two eye-like depressions. The author did not give the size of the fruit. Tripathi, Mishra, and Sharma (1999) recovered a *Cocos*-like petrified fruit from the Tertiary (66–5 Mya) of Amarkantak, Madhya Pradesh. The fruit measured 13 × 10 × 6 cm, and was oval in shape. Its third ridge was ill-defined. The epicarp showed fine longitudinal striations of fibers. The central cavity was 6 × 4 cm in size. The fruit wall was 0.6–1.5 cm thick, mesocarp 2–5 mm thick and consisted of fibers, and endocarp, 0.4–1.5 mm thick, and was heterogeneous (sic). The fruit dimensions are well within the fruit size limits of extant coconut fruits from South and Southeast Asia, the authors claimed.

Shukla, Mehrotra, and Guleria (2012) studied in detail *Cocos sahnii*, which Kaul (1951) had described earlier from Khapurdi. This region is now a part of the Thar Desert. The site is located in Khapurdi Formation. It is the youngest Tertiary deposit in the Barmer region, Rajasthan. It consists of fuller's earth (a type of clay used for shrinking cloth) with carbonaceous streaks with gypseous clay in lower part. The authors assigned it to Early Eocene (Eocene = 57–36 Mya). They studied two *Cocos*-like specimens: one, an impression of a mesocarp, and the other an endocarp with fibrous covering. The latter specimen measured 11.7 × 5.8 cm, was ovoid and assymetrical in shape. In the former one, the mesocarp impression consisted of parallel, longitudinally oriented fibers. The endocarp showed two micropyles and two longitudinally oriented ridges. The associated flora consisted of tropical evergreen trees such as *Calophyllum inophyllum*, *Messua ferrere*, and *Garcinia* spp. All of them are known strand- and water-loving plants. They found also remains of crabs, fishes, turtles, and gastropods (Lakhanpal, 1970; Lakhanpal & Bose, 1951). These associations suggested a lacustrine or lagoonal environment (Rana et al., 2006).

Prasad, Khara, and Singh (2013) observed *Cocos*-like leaf and fruit impressions from Deccan intertrappean beds of Mohgaokalan, Chhindwara district, MP (22°1′N, 79°11′E). According to the authors, this is one of the two localities in MP having the richest fossiliferous intertrappean deposits. Coconut was one of 10 plant species found in this locality. The authors found two forms of *Cocos* plants. They named the two forms, *C. palaeonucifera* and *Cocos nucifera*. They assigned them to early Tertiary (Maastrichtian–Danian, 70–62 Mya). Generally, most samples were in petrified state, and only few were fossils. The petrified samples consisted of 5–6 immature fruits, oval to triangular in shape with broken epicarp and mesocarp. The mesocarp impressions measured 1.8 × 1.8, 1.5 × 1.5 cm, and 2.0 × 1.5 cm in size. The *Cocos nucifera* forms were in a petrified state.

Recently, Srivastava and Srivastava (2014) discovered a *Cocos*-like specimen in Binori Reserve Forest, Ghansor, Seoni district, MP (22°N, 79°E) in sediment material of approximately Maastrichtian–Danian (70–61 Mya) period. This consisted of Deccan intertrappean sediments deposited in a lacustrine and fluvial environment. During this period, India had been separated from the rest of Gondwana, but had not yet collided with Asia. They named it *Cocos binoriensis*. It consisted of a three-dimensionally preserved drupe, ovoid in shape, and showed clear longitudinal ridges. The exocarp was smooth, mesocarp was fibrous with vertical and horizontal [sic] fibers on the inner surface of endocarp. The authors have claimed this as the oldest fossil specimen of *Cocos nucifera* (Figs. 4.3 and 4.4).

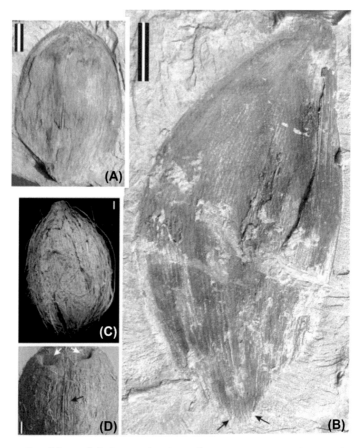

Figure 4.3 Fossil fruit of cf. *Cocos sahnii* (actual size 5.5 cm). (A) Endocarp of the fossil showing two eyes. (B) Close-up of the mesocarp of the fossil fruit showing shape and longitudinal fibers. (C) Look-alike of (B) of the present-day coconut with mesophil. (D) Endocarp of present-day coconut showing two eyes.

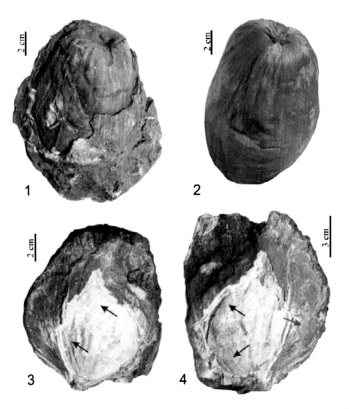

Figure 4.4 Fossil fruit of cf. *Cocos binoriensis* (actual size, 11.7 × 10.0 cm). 1. Showing shape, size, and two longitudinal ridges. 2. Present-day coconut fruit. 3. Longitudinal broken part of the fossil showing fibrous mesocarp. 4. Counterpart of (3) showing endocarp inside view. *Reproduced from Srivastava, R., & Srivastava, G. (2014). Fossil fruit of Cocos L. (Arecaceae) from Mastrichian – Damian sediments of central India and its phytogeographical significance.* Acta Palaeobotanica, *54, 67–75 with permission.*

Dransfield et al. (2008) had commented that the fossil record of palms in the Indian subcontinent has been very rich. They belonged to the Tertiary period (66–9 Mya) and represented possibly at least 17 extant genera.

Most of the pollen records have been obtained from the Deccan traps, the western and eastern peninsular Indian coast, and peninsular India, in the context of oil explorations (i.e., as pollen). Incidentally, it appears that only very preliminary types of studies have been done on them (Harley & Baker, 2001; Rao & Ramanujam, 1975; Singh & Rao, 1990).

Only two *Cocos*-like fossils have been reported from regions other than New Zealand and India, Queensland, Australia, and Colombia, South America.

4. FOSSILS FROM OTHER REGIONS

There are two such reports. First, in Australia, Rigby (1995) recovered permineralized *Cocos*-like fruits from Chinchilla Sand (26°47′S, 150°37′E), 5 km from Chinchilla on Fairmeadow Road, where it crossed a gully. The location was at 300-m elevation and c. 250 km inland from Brisbane, off the Pacific Ocean. The age of the Sand appeared to be latest Pliocene, or possibly, basal Pleistocene. The fruit showed evidence of some hulling. It measured 10.0×9.5 cm, which the author observed was "within the size range of living coconuts." This region is known for the presence of fossils of crocodiles, freshwater tortoises, birds, and marsupials (Rigby, 1995) (Fig. 4.5).

The second *Cocos*-like fossil reported from outside New Zealand and India was from a rain-forest site in northeast Colombia, South America (Gomez-Navarro, Jaramilo, Herrera, Wing, & Callejas, 2009). The authors recovered 12 specimens from an open-cast coal mine between the coal seams 125 and 130. They belonged to five species from a Middle to Late Paleocene Cerrejon Formation outcropping in the Rancheria River valley in northern Colombia (Tobacco High Dippit; 11°62′N, 73°32′W).

Its age and stratigraphy were Middle to Late Paleocene (Paleocene: 67–57 Mya) based on pollen carbon isotope studies. The specimens consisted of inflorescences, fruits, and leaves. They were found to relate to three of the five lineages of Arecaceae, the monotypic *Nypa*, Calamideae, and Coryphoideae. The inflorescence and a single fruit belonged to the subfamily Arecoideae, and the fruit was assignable to the genus *Cocos*. This fruit was compressed in appearance, ovoid in shape, and measured 250×150 mm in size, but had only 4–8 mm thickness (Figs. 4.6 and 4.7). The apex was acute, asymmetrically located with an inconspicuous, longitudinally oriented ridge. Due to the compaction of the fruit, its internal structure was not distinguishable, and hence, it was not possible to determine if it had the three pores—which is characteristic of tribe Cocoseae—or not. The authors concluded that "among the modern genera of subtribe Attaleinae, only *Cocos* is similar to the fossil in being large and having inconspicuous ridges."

5. PALEOPALYNOLOGY

Palynology is the study of pollen grains and pores. Only very limited studies have been done on *Cocos* palynology.

Thanikaimoni (1966, 1970) and Sowunmi (1972) have described the morphology of the coconut pollen. The pollen morphology of *Rhopalostylis* (Tribe Araceae) and *Jubaea* (Tribe Cocoseae) are similar to that of the

Figure 4.5 *Cocos*-like fossil recovered from Chinchilla Sand, Qld., Australia (actual size, 10.0 × 9.5 cm). *Reproduced from Rigby, J. F. (1995). A fossil* Cocos nucifera L. *fruit from the latest Pliocene of Queensland, Australia. In D. D. Pant (Ed.), Birbal Sahni Centennial Volume (pp. 379–381). Allahabad, India: Allahabad University, Allahabad, India/South Asian Publishers with permission.*

coconut. Sowunmi (1968, 1972) indicated also that in addition to *Jubaea, Maxmiliana, Orbignya, Butia, Astrocaryum, Jubaeopsis, Sheelea, and Parasheelea* also possess *Cocos*-type pollen. Incidentally, all these genera belong to the subtribe Attaleinae to which belongs also the genus *Cocos*.

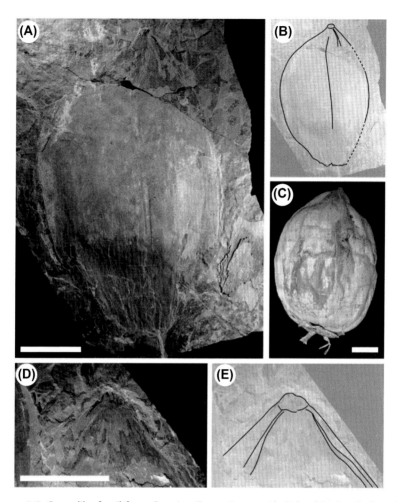

Figure 4.6 *Cocos*-like fossil from Cerrejon Formation, north Colombia, South America (actual size, 25.0×15.0 cm). (A) Whole fruit of cf. *Cocos* fruit. (B) Drawing of whole fruit. (C) Present-day coconut, *C. nucifera*. (D) Counterpart of *Cocos* sp. Apex. (E) Drawing showing longer axis. *Reproduced from Gomez-Navarro, G., Jaramilo, C., Herrera, F., Wing, S. L., & Callejas, R. 2009. Palms (Arecaceae) from a Palaeocene rainforest of northern Columbia. American Journal of Botany, 96, 1300–1312 with permission.*

Cocos-type pollen has been observed in several fossiliferous sediments in the Philippines, Thailand, India, and Sri Lanka (Maloney, 1993; Ramanujam, 2004). However, its presence was not seen continuously anywhere from the Miocene (23–5 Mya) to the present time. The oldest Quaternary (1.6–0.01 Mya) sediments from Asia covered 300,000 years (Van der Kaars, 1990). They came from the deep-sea cores taken from the Molucca Sea, Kau Bay, Ceram Trench, Lombok Ridge, and Argo Abyssal Plain, all from Southeast Asia.

Figure 4.7 Coconut endocarp fragments recovered from Anawau swamp, Anietyum Island, Vanuatu (Period 5420 ± 90 BP). *Reproduced from Spriggs, M. (1984). Early coconut remains from the South Pacific. Journal of the Polynesian Society, 93, 71–76 with permission.*

The earliest *Cocos* pollen from the Holocene was obtained from Te Roto (Cook Islands, 17°0′S, 160°0′W, 1350 km east of Fiji) c. 8600 Ya and Atiu (also, Cook Islands, c. 7800 Ya) (Parkes & Flenley, 1990). However, the pollen content was very low. This was a period when present-day humans (*Homo sapiens*) were not known to have reached this area (Bellwood, 2013). The authors proposed that a species of the *Cocos* group was present in the western Pacific 8000 Ya as part of the natural population.

A dramatic increase in pollen took place during 550–600 Ya. The upper sediments from Lake Temae, Moorea, Society Islands (17°0′S, 151°0′W; French Polynesia) contained coconut pollen (Parkes & Flenley, 1990), which predated to 1500 BP at its oldest, but they were not found from the 500-year old record of Lake Vaihiria, Tahiti (French Polynesia, 17°37′S 149°29′W). The author reasoned that this might have been because of the high latitude of the site, 17°S.

The oldest report from Southeast Asia of coconut pollen was obtained from a core extracted close to the archeological site of Khok Phanom Di, central Thailand (2000–1500 BP). But it consisted of only a single pollen and it was not dated either. Hence, it is not seriously considered.

Maloney (1993) concluded that in Southeast Asia, coconut might have been present from 6000 BP, but it was not possible to know if they were domesticated or wild coconuts. As for the Pacific Ocean, the Cook Islands had not been occupied by present-day humans 8600Ya, and hence, the pollen that was recovered from there could have been of naturally occurring coconut palms, or alternately, brought there by winds.

McCormack (2005) found coconut palm pollen in the Cook Islands (17°S 160°W; 17°S 160°W) in lake sediments dated to 6600 BCE in Atiu and dated to 5300 BCE in Mangaia. The earliest direct evidence of Polynessian settlement in southern Cook Islands has been from Rarotonga on Motu Tapu around 900 CE (Bellwood, 2013). Hence, the coconut pollen could have been of either wind-borne pollen, or from natural stands of coconuts that were growing in the islands at that time.

6. ARCHEOBOTANY

6.1 Introduction

Compared to other crop plants having comparable wide distribution as the coconuts, the number of archeological findings obtained in the coconut has been few. This is despite the fact that most components of the palm—hard stem, thick hard fruits, and the nut, high silica-containing leaves, etc.—are comparatively resistant to decomposition.

One main reason for this could be that most regions, where the coconuts are grown, have not been studied sufficiently. A second reason could be the submergence of the coastal areas of South and Southeast Asian region following the sea level rise (up to 90 m, Clark et al., 2012) after the Last Glacial Maximum (24,000–19,000Ya). There are several reports of this also having some level of influence in the western Pacific region.

Marcotte-Rios and Bernal (2001) had reviewed palm archeology in the New World. They did not find here any record of *Cocos*. But remains of certain near-related genera such as *Acrocomia*, *Attalea* sensu lato, *Astrocaryum*, *Bactris*, *Syagrus*, and *Elaeis*—all belonging to the *Cocos* alliance—and *Oenocarpus*, have been predominant. Members of all these genera are edible and oil yielding, and the indigenous people have been using them for various purposes. The remains included carbonized or dry endocarps or entire seeds, phytoliths, pollen, and implements. No *Cocos* has been recovered from any of the New World archeological remains till date.

Fosberg and Corwin (1958) observed the impression of a strongly plicate single bent section of a seedling leaf in 1954 at Pagan, Mariana Islands (Micronesia, 18°01′N, 145°41′E, 430 km north of Guam, 1700 km southsoutheast of Tokyo). The authors assigned the fossil to late Quaternary age.

Northern Marianas is a chain of volcanic summits, c. 450 km long. Their maximum elevation above sea level is c. 800 m and minimum c. 1800 m below sea level (Wikipedia, 2015). Presently, the coconuts grow abundantly in these islands. Most of them consist of planted crop.

Spriggs (1984) studied a swamp site in Ancilum Island, Vanuatu (20°14′S, 169°46′E). It is the southernmost inhabited island of Vanuatu. The Anawau swamp site was found near Ancleauhat village. It was 15 ha in area and 80-m wide. It was located behind a sand bar. The author found at 175 cm depth numerous pieces of coconut roots (layer # 6) in the bottommost layer (5040 ± 370 BP) and endocarp from layers # 5 and 4 (5420 ± 90 BP and 5410 ± 100 BP). The largest endocarp fragments, including parts of the intercarpellary ridge and basal eyes, were less than 35 mm in length and 20 mm in width. The fragments measured 18 × 27 mm. Spriggs felt that this finding likely appeared to fall within Harries' niu kafa type. Stabilization of sea level in this region likely occurred 6000–5000 Ya. Earlier research (Allen, 1980) had failed to find any convincing evidence for human occupation southeast of Bismarck Archipelago prior to the Lapita expansion of 3500–2500 BP. Hence, this finding may indicate also the presence of coconuts as a component of the natural vegetation in Vanuatu during this period.

Spriggs observed that the Ancilum discovery of coconut was the earliest archeological record of the coconut in the Pacific Ocean. It continues to be so even now (Carson, 2012).

The next earliest record was from Aitape in northeast Papua New Guinea at 4555 ± 80 BP. It was associated with human skeletal remains (Hossfeld, 1965).

Kirch and Yen (1982) obtained charcoal and coconut remains from Tikopia, southeast Solomon Islands dated to 3360 ± 130 BP and 2695 ± 90 BP. They were obtained in the context of the presence of coconuts as part of strand vegetation in Papua New Guinea.

Lepofsky, Harries, and Kellum (1992) reported about the occurrence of early forms of cultivated coconuts from Mo'orea Island, French Polynesia (Society Islands, c. 25 km west of Tahiti, 20°S, 145°W) (Fig. 4.8). The authors (DL and MK) recovered several anaerobically preserved whole coconuts from Mo'orea Island, from a stream bank on the extensive alluvial flat of Opunohu valley. They were found 1.5 km inland of the present Opunohu Bay, at a depth of 2.3 m below the then ground surface and approximately 5 m above the current sea level. They recovered entire nuts, but none of them contained any endosperm. The nuts matched neither the extreme wild nor domesticated form of coconut. They showed intermediate

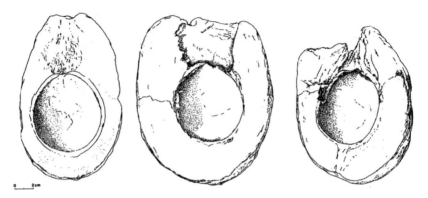

Figure 4.8 Anaerobically whole coconuts from Mo'orea, Society Islands, French Polynesia (Period 1360±60 BP). *Reproduced from Lepofsky, D., Harries, H. C., & Kellum, M. (1992) Early coconuts in Mo'orea, French Polynesia. Journal of the Polynesian Society, 101, 299–308 with permission.*

characteristics of wild and domesticated forms. They had relatively thick husk, oblong shape, and moderately thick shell. Two nuts were carbon dated. Their adjusted dates were 1270±60 BP and 1360±BP. The lowest layer, where some inaerobic coconuts (and floral parts) were recovered, consisted of waterlogged homogenous black silts of about 45-cm thickness. Society Islands are thought to have been first occupied by humans c. 650 BP ago (Bellwood, 2013). Thus, these nuts could have come as only drift nuts, or more possibly, from the coconuts occurring in the Islands as part of the natural vegetation.

We can thus see that the western and central Pacific region used to have the coconut growing as part of the natural vegetation from at least 8000 BP (Table 8.3).

CHAPTER 5

Phylogeny

1. INTRODUCTION

Phylogenetics is the study of the relationships of groups of organisms as reflected by their evolutionary history (King, Stansfield, & Mulligan, 2006).

In the last 50 years, systematic biology has been literally transformed by phylogenetic studies relationships are now being based increasingly on the results of the analyses of clearly formulated data sets of characters. This is very evident from the meta-analyses of angiosperm taxonomy (Angiosperm Phylogeny Group III (APG III), 2009). Palm classification too has benefited substantially by phylogenetic studies (cf. Dransfield et al., 2008). The consequences of this development in understanding coconut systematics and relationships will be evident also from the accounts given in the following pages.

The cultivated coconut, *Cocos nucifera* L. as we have noted before, belongs to the genus *Cocos* L. (the only species in the genus), subtribe Attaleinae Drude (*Cocos*, one of 11 genera), (one of three subtribes, the other two, Bactridinae and Elaeidinae), tribe Cocoseae Mart. (Cocoseae, one of 14 tribes), subfamily Arecoideae one of five subfamilies, family Arecacae Brecht. and J. Presl. (known also as Palmae Juss.), order Arecales Bromhead (the only family in Arecaceae), division Monocotyledonae, and class Magnoliopsida. The order Arecales is one of five orders of a group termed commelinids under class Magnoliopsida (Angiospermae, including Liliopsida). The Monocotyledonae includes 86 families arranged in 11 orders. There are, in all, 413 recognized families, referable to 11 orders (APG III, 2009; Mabberley, 2008).

Traditionally, the palm family had been linked by several authors to three other monocot families, Araceae, Cyclanthaceae, and Pandanaceae (Dransfield et al., 2008). Nevertheless, by using molecular phylogeny, the palms are now seen to resolve better within a major group of families termed commelined monocots. They include four orders and one unplaced family: Arecales, Commelinales, Poales, Zingiberales, and the family Dasypogonaceae. However, the relationships among them are still unresolved (APG III, 2009).

The Coconut
ISBN 978-0-12-809778-6
http://dx.doi.org/10.1016/B978-0-12-809778-6.00005-X

The latest classification of the Family has been done by Dransfield et al. (2008) in Genera Palmarum II (GP II). It recognizes 2400 plus species and 183 genera in 14 tribes, under five subfamilies. A unique feature of the family is the occurrence of a large number of monospecific genera—56/183 = 31%. There are also as many as 94/183 genera, i.e., 51% genera that have three or fewer species. The family is characterized by relatively high levels of endemism (a species that is native to a region or place; not introduced; King et al., 2006), variable habits, and homoplasy (structural similarity in organisms not due directly to inheritance from a common ancestor or development from a common embryonic primordium; King et al., 2006).

2. MOORE'S CLASSIFICATION

The last classification of palms based on morphological characters was done by Moore (1973). However, he avoided using classical taxonomic categories below the family level.

Moore (1973) classified the Family into 15 groups under five "lines," but without assigning any rank. The group to which the coconut belonged was called Cocosoid palms, under the Arecoid Line. Moore distinguished the Cocosoid palms from the "sister" group, Arecoid palms, by the presence in the former of thick, bony endocarp enclosing one to three, or rarely more, seeds and with three, or rarely more, pores at, below, or above the middle (vs. fruits with a thin or rarely thick endocarp with endocarp and lacking pores in Arecoid palms). The Cocosoid palms consisted of c. 28 genera and 583 species.

Moore then divided the Cocosoid palms into three units/alliances, Cocos alliance, Elaeis alliance, and Bactris alliance. The Cocos alliance with 20 species was further divided into seven units: (1) Cocos unit (*Cocos* only); (2) Butia unit (*Butia* only); (3) Jubaea unit (*Jubaea, Jubaeopsis*); (4) Syagrus unit (*Syagrus, Arecastrum, Barbosa, Rhyticocos, Chrysalidosperma, Arikuryroba, Microcoelum,* and *Lytocaryum*); (5) Parajubaea unit (*Parajubaea*); (6) Allagoptera unit (*Allagoptera* and *Polyandrococos*); and (7) Attalea unit (*Attalea, Scheelia, Parascheelia, Orbignia,* and *Maxmiliana*): total 20 genera.

The innate soundness of Moore's (1973) classification has been confirmed by subsequent phylogenetic and taxonomic studies.

Out of the cited 20 genera, GP II (2008) has recognized nine genera as valid ones—*Cocos, Butia, Jubaea, Jubaeopsis, Syagrus, Lytocaryum, Parajubaea, Allagoptera,* and *Attalea.* The remaining 11 genera were considered synonyms under:

Syagus: *Arecastrum, Arikuriroba, Rhyticocos, Barbosa, Chrysalidosperma*
Jubaea: *Microcoelum*

Allagoptera: *Polyandrococos*
Attalea: *Scheelia, Parascheelia, Orbignya, Maxmiliana*

GP II (2008) has included the following 11 genera in the tribe Cocoseae: *Beccariophoenix* (2 species), *Jubaeopsis* (1 species), *Voaniola* (1 species), *Allagoptera* (5 species), *Attalea* (c. 69 species), *Butia* (9 species), *Cocos* (1 species), *Jubaea* (1 species), *Lytocaryum* (2 species), *Syagrus* (31 species), and *Parajubaea* (3 species); total: c. 125 species.

Moore had recognized Cocosoid palms as a single group because of their distinctive long endocarps with three pores. He recognized three sub-group alliances within the cocosoids, the Bactris, Cocos, and Elaeis alliances. Uhl and Dransfield (1987) has now recognized Moore's Cocosoid palms into tribe Cocoseae as one among 14 tribes of the subfamily Arecoideae with 18 genera and c. 513 species grouped into three subtribes. The three subtribes are: Attaleinae (11 genera, c. 125 species, including *Cocos*), Bactridinae (5 genera, c. 385 species), and Elaeidinae (2 genera, 3 species). Members of Attaleinae are unarmed, whereas those of the other two subtribes are armed.

Readers are invited to refer to Table 6 (pp. 134–135) in GP II (2008) to see the historical evolution of palm classification beginning Martius, 1849–53. Excerpts from the classification of Cocosoid palms are given in Fig. 3.2A and B.

3. PHYLOGENY OF THE COCOSOID/COCOSEAE PALMS

This has been studied by several authors using a range of DNA molecular techniques beginning with Uhl et al. (1995). The most detailed studies of the cocosoid palms have been done by Baker et al. (2009, 2011) and Meerow et al. (2009, 2014).

In the following pages, we shall consider in particular the studies done on the tribe Cocoseae/Cocosoid Line (*sensu* Moore) with special reference to the genus *Cocos*. The early studies understandably were meant to determine the suitability of various DNA regions for phylogenetic analysis.

One of the earliest studies was of Lewis and Doyle (2001a, 2001b). In the 2001 study, the authors used the nuclear gene, malate synthase, in 29 palm genera with three outgroup taxa. This included three genera of Cocoseae, *Allagoptera, Bactris*, and *Baccariophoenix*. Nevertheless, the resolution was unsatisfactory. Lewis and Doyle (2001a, 2001b) studied the phylogeny of the then-largest tribe of the palm family, Arecaceae (178/189 genera), using two low-copy nuclear genes, *MS* and *PRK*. They used 54 palm species, which included representatives of all of the then-extant 17 subtribes. This included also 2/20 species of the subtribe Cocoseae, *Beccariophoenix madagascariensis* and *Allagoptera arenaria*. The results of the combined analysis showed that a clade of Indo-Pacific taxa was resolved with strong support. Further, Arecaceae was resolved as polyphyletic, as in the authors' 2001 study.

Hahn (2002a, 2002b) analyzed the phylogeny of the Arecoid Line (*sensu* Moore) using 65 genera and five DNA sequences and two plastid primers—nuclear DNA (18 S) and chloroplast DNA (*atp*B, *rbc*L, *rps* 16, and *ndh* F, and *trn* L-*trn* F). The order Butiinae (in which *Cocos* was placed at that time) was represented by six species—*Allagoptera arenaria*, *Butia eriospatha*, *Cocos nucifera*, *Lytocaryum weddelianum*, *Syagrus glaucessans*, and *Voaniola girardii*. The author estimated the phylogenetic relationships using maximum parsimony (MP) (see Figs. 5.1 and 5.2) and maximum likelihood (ML) methods.

In Hahn's (2002b) study, the order Cocoeae was resolved as monophyletic, but with only a weak support, 62% BS (bootstrap proportions). Subtribe Beccariophoenicinae was sister to the nonspiny Cocoeae (61% BS).

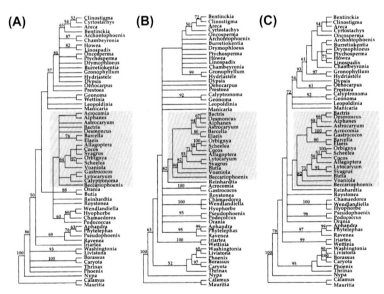

Fig. 1. Strict consensus trees of maximum-parsimony analysis on sequence data with simple sequence data removed. (A) Consensus of 25,000 MP trees from analysis of 1958 bp of noncoding data (L = 629, CI = 0.83, RI = 0.74). (B) Consensus of 12,019 MP trees from analysis of 5038 bp of coding data (L = 1150, CI = 0.64, RI = 0.62). (C) Consensus of 2016 MP trees from analysis of 6996 bp of combined coding and noncoding data (L = 1705, CI = 0.69, RI = 0.62). Bootstrap support values > 50% are indicated above branches.

Figure 5.1 Strict consensus of maximum-parsimony analysis on sequence data with simple sequence data removed. *Reproduced with permission from Hahn, W. J. (2002a). A phylogenetic analysis of the Arecoid line of palms based on plastid DNA sequence data. Molecular Phylogenetics and Evolution, 23, 189–204.*

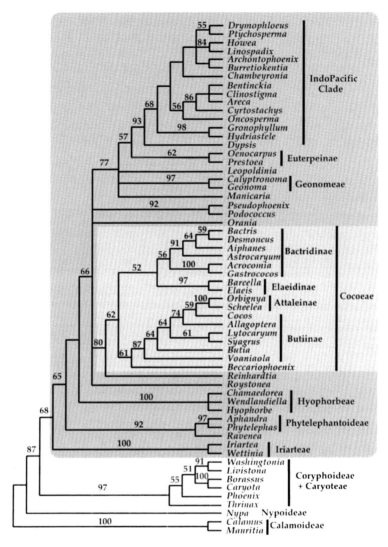

Fig. 2. Strict consensus of 480 maximum-parsimony trees from the combined coding and noncoding cpDNA data set with binary indel characters included. The Arecoid Line is in the shaded area. Length = 1848; CI = 0.69; RI = 0.62. Bootstrap support values > 50% are indicated above branches.

Figure 5.2 Strict consensus of 480 maximum-parsimony trees from combined coding and noncoding chloroplast DNA (cpDNA) data set. More details beneath the figure. Cocoseae are in the *light-shaded portion. Modified with permission from Hahn, W. J. (2002a). A phylogenetic analysis of the Arecoid line of palms based on plastid DNA sequence data.* Molecular Phylogenetics and Evolution, 23, 189–204.

Subtribes Attaleinae and Butiinae formed a well-supported clade (87% BS). The author then argued that though Asmussen and Chase (2001) and Hahn (2002a) had also got similar results, the large number of morphological characters that identified the group could be taken to support monophyly of Cocoeae. The author also highlighted the sister relationship of *Reinhardtia* to the Cocoeae obtained in this study. Hahn found further that the five subtribes of Cocoeae were resolved as two major clades: (1) subtribes Bactridinae + Elaeidinae, with low molecular support (52% BS), but with strong support of morphological characters (incidentally, they are both armed); and (2) subtribes Attaleinae, Beccariophonecinae, and Butiinae. Their significant morphological features were endocarp pores below the equator, lack of spines, and pistillate flowers borne superficially. This group was moderately supported in MP trees (61% BS), but with *Beccariophoenix* excluded. Butiinae was paraphyletic with Attaleinae embedded in it.

As for the genus *Cocos* (then, in subtribe Butiinae; now in subtribe Attaleinae), the MP analysis showed a sister relationship with a clade containing *Orbignya* and *Scheelia* (now, both genera are included as synonyms of Attaleinae; Dransfield et al., 2008). A similar result for *Cocos* was obtained in the ML analysis (see Fig. 5.3) from coding and noncoding cp DNA tests also.

Gunn (2004) studied the phylogeny of Cocoeae (*sensu* Moore, 1973) with 35 species using the nuclear *PRK* gene. The author sampled all its then five subtribes along with two outgroups and one in-group. From the coconut, Gunn included three varieties, two talls and one dwarf. She analyzed the molecular data using MP analysis. The MP topology of all the clones showed monophyly of the cloned taxa. The Cocoeae divided itself into two main clades, the nonspiny (94% BS) and spiny (98% BS) species. *Barcella odora* was basal to Cocoeae (with a low 55% BS), with rest of the clade having 78% BS. The latter (nonspiny) clade further divided itself into two clades comprising *Cocos* and *Attalea* alliances, with *Jubaeopsis caffra* remaining unresolved. The *Cocos* alliance had 53% BS, and the *Allalea* alliance, 100% BS (Figs. 5.4–5.6).

In Gunn's study, the Bayesian analyses recovered trees that were consistent with the MP topology. The author felt that the closest relative to *Cocos nucifera* probably belonged to the *Cocos* alliance [*Allagoptera + Polyandrococcus* (now, *Allagoptera*, in GP II), *Butia + Jubaea*, *Syagrus smithi + S. amara*, and *Parajubaea*], even though, all of them now occupy habitats with low rainfall and sandy soils. The three *C. nucifera* varieties formed a monophyletic group (100% BS), but it was subtended by a moderately long branch. The author then surmised that the most closely related taxon to *C. nucifera* was

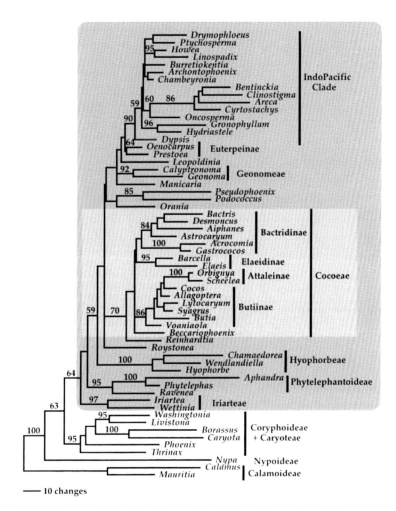

Fig. 3. Maximum-likelihood tree from combined coding and noncoding cpDNA
data set. The Arecoid Line is in the shaded area. ML bootstrap support values > 50%
are indicated above branches. − log likelihood = 22280.09. Substitution model
parameters: AC = 2.001; AG = 3.205; AT = 0.620; CG = 1.015; CT = 2.943; GT = 1;
PInvar = 0.512; shape parameter alpha = 0.623.

Figure 5.3 Maximum-likelihood tree from combined coding and noncoding chloroplast DNA (cpDNA) data set. Arecoid in *shaded area* with subtribe Attaleinae in *light-
shaded portion. Adapted with permission from Hahn, W. J. (2002a). A phylogenetic analysis
of the Arecoid line of palms based on plastid DNA sequence data.* Molecular Phylogenetics
and Evolution, 23, 189–204.

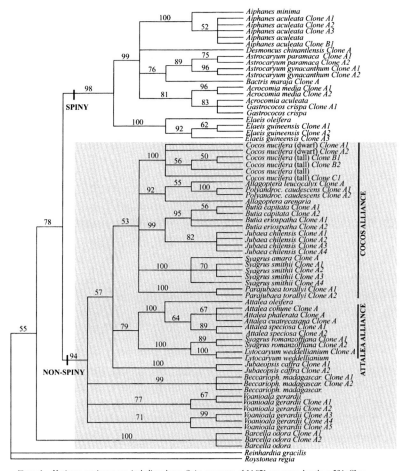

Figure 1. Maximum parsimony tree including clones. Strict consensus of 14,071 trees; tree length = 524; CI = 0.733; RC = 0.668; RI = 0.911. Percentages above branches show bootstrap values. The abbreviation *Beccarioph. madagascr.* = *Beccariophoenix madagascariensis.*

Figure 5.4 Strict consensus of 14,071 maximum-parsimony trees subtribe Attealeinae in *shaded portion. Adapted with permission from Gunn, B. F. (2004). Phylogeny of the Cocoeae (Arecaccae) with emphasis on* Cocos nucifera. Annals of the Missouri Botanical Garden, 91, *505–525.*

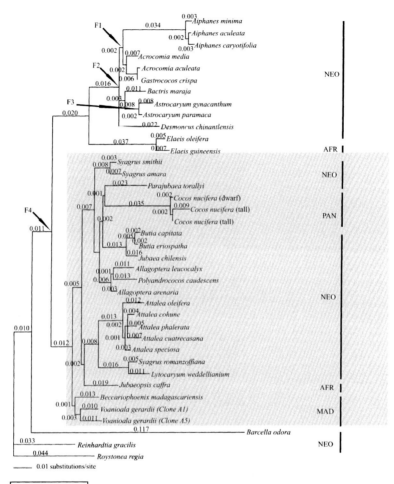

Figure 2.   Maximum likelihood tree. Model used = HKY + G; base frequencies: A = 0.2948, C = 0.1990, G = 0.2256, T = 0.2806. Transition/transversion ratio = 1.8755; gamma distribution shape = 1.0177; Aikaike Information Criteria (AIC) = 7670.2085. Values above branches show number of nucleotide substitutions/site. F1, F2, and F3 = Cocoeae fossils. F1 = Wharekuri, New Zealand (32 mya). F2 = Boulder Hill, New Zealand (38–45 mya). F3 = Waihao beaked, New Zealand (27–31 mya). F4 = *Cocos intertrappeansis*, India (50 mya).

Figure 5.5 Strict consensus of maximum-likelihood (ML) trees Attaleinae in *shaded portion*. More details given earlier. *Adapted with permission from Gunn, B. F. (2004). Phylogeny of the Cocoeae (Arecaccae) with emphasis on* Cocos nucifera. Annals of the Missouri Botanical Garden, *91, 505–525.*

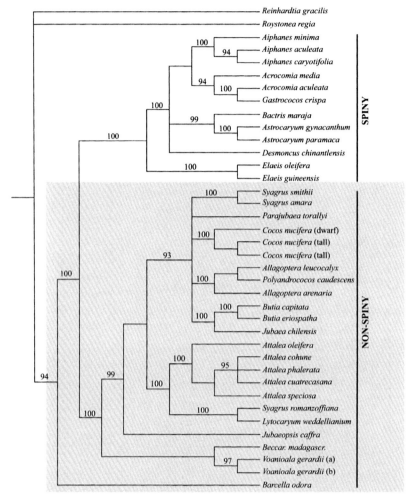

Figure 3.   Bayesian Tree. Values above branches show percentages of prior probability. The abbreviation *Beccar. madagascr.* = *Beccariophoenix madagascariensis.*

Figure 5.6 Bayesian Tree. *Reproduced with permission from Gunn, B. F. (2004). Phylogeny of the Cocoeae (Arecaccae) with emphasis on* Cocos nucifera. Annals of the Missouri Botanical Garden, 91, 505–525.

Parajubaea, because in ML analysis, *P. torallyi* resolved as sister to *C. nucifera.* This species, *P. torallyi*, grows in the humid valleys of sandstone mountains of the central and southern Bolivian Andes at 2000-m altitude.

Asmussen et al. (2006) used plastid DNA region *matK* to sequence 178 palm species and 10 commelinid monocot outgroup species (Fig. 5.7). They

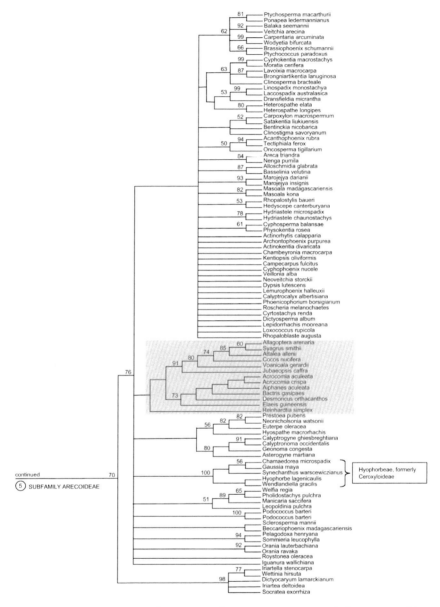

Figure 5.7 Strict consensus of 30,000 equally most parsimonious cladograms resulting from Fitch parsimony analyses of combined *matk*, *rbcL*, *rps16* intron and *trnL-trnF* data sets. Bootstrap percentages for the clades are given above the branches. Five clades are labeled 1–5 for the discussion of a new subfamily classification in the text. The clades corresponding to the five subfamilies of the new classification are labeled with the subfamily name. *Adapted and Reproduced with permission from Asmussen, C. B., Dransfield, J., Deickman, V., Barford, A. S., Pinaud, C., & Baker, W. J. (2006). A new subfamily classification of the palm family (Arecaceae): evidence from plastid DNA phylogeny.* Botanical Journal of the Linnean Society, 151, *15–38.*

combined this with part-new and part-published plastid DNA sequences of *trnL-trnF*, *rps*16 intron, and *rbcL*. The results showed foremost that critical basal nodes were fully resolved. Further, of interest to our study, the subfamily Arecoideae was modified to exclude the tribe Caryoteae and include tribe Hyophorbeae (now, synonym of tribe Chamaedoreae). The species *Cocos nucifera* showed a sister relationship with a clade containing *Attalea alleni*, which, in turn, was sister to a clade containing *Allagoptera arenaria* and *Syagrus smithii*.

Meerow et al. (2009) carried out a detailed phylogenetic analysis of subtribe Attaleinae, tribe Cocoseae, using material which included, till then the most intensive sampling of the subtribe Attaleinae used for a molecular study. They used 75 taxa—72 from all the genera of Attaleinae and three from outgroups. The coconut was represented by six accessions, Atlantic Tall, Green Malayan, Red Malayan, Red Spicata, Panama Tall, and Niu Leka. The authors employed DNA sequences of seven *WRKY* transcription factor loci. They analyzed the data with maximum parsimony (MP), ML, and Bayesian approaches (BA). The analyses produced highly congruent and well-resolved trees. They felt that the occasional incongruence was more due to insufficient resolution at various nodes within each locus. The ML tree was essentially identical to the MP tree, but the terminals were more fully resolved in the former. The BA of the combined data matrix was also congruent with MP and ML (except for *Bactris*) (Fig. 5.8). With combined analysis, the sequence matrix consisted of 5831 total characters of which 974 characters (17%) were parsimony informative. Combining all seven *WRKY* loci yielded the most fully resolved trees and the highest bootstrap support with both MP and ML.

The combined *WRKY* data matrix of Meerow et al. (2009) supported monophyly of all the genera of the subtribe Attaleinae, except of *Syagrus*, which was paraphyletic with *Lytocaryum* (see Fig. 5.9). The crown node consisted of three clades: (1) the African genera (91/95% BP), in which *Baccariophoenix* was sister to a *Jubeopsis/Voaniola* clade (84/84% BP); (2) the American clade, which showed a sister relationship with the African clade; and (3) the American clade resolving as two monophyletic groups. The better supported of the two groups (95/90% BP) showed that *Cocos* was strongly supported as sister to *Syagrus,* in which was embedded *Lytocaryum*. (4) The second less well-supported clade (72/96% BP) united *Butia* and *Jubaea* into a clade (100/100% BP) that was sister to a fairly well-supported clade (88/96% BP) of *Allagoptera, Parajubaea,* and *Polyandrococcus;* and (5) a monophyletic *Attalea* was sister to the rest of the clade.

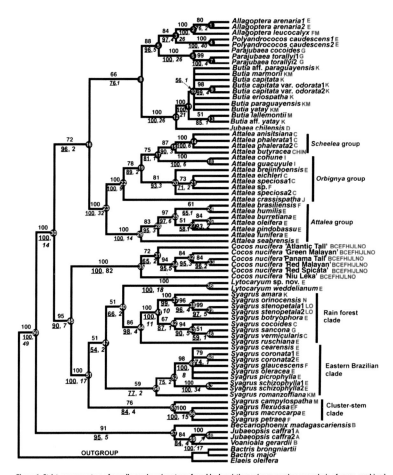

Figure 1. Strict consensus tree of equally parsimonious trees found by heuristic maximum parsimony analysis of seven combined WRKY loci sequences aligned across Areceaceae tribe Cocoseae subtribe Attaleinae. Numbers above branches are bootstrap support percentages. Numbers below branches are ML bootstrap support (underlined) and non-partitioned decay indices (italic). The numbers at each node refer to Table 2, which see for partitioned decay indices. Letter designations in red are area distributions of terminal taxa: A = South Africa, B = Madagascar, C = Amazonas (north, east, south), D = Chile, E = Eastern Brazil, F = Central Brazil, G = Andes, H = Central America, I = Mexico, J = West Indies, K = Southern Brazil, L = Northern South America, i.e., Caribbean coastal Venezuela and Colombia, French Guiana, Guyana, Surinam, M = Argentina-Paraguay-Uruguay, N = Colombia-Venezuela (*llanos* region), and O = western Amazonas.
doi:10.1371/journal.pone.0007353.g001

Figure 5.8 Strict consensus tree of equally parsimonious trees found by heuristic maximum-parsimony analysis of seven combined WRKY loci sequences aligned across subtribe Attaleinae. Detailed explanations given beneath figure. *Reproduced from Meerow, A. W., Noblick, L., Borrone, J. W., Couvreur, T. L. P., Mauro-Herrera, M., Kuhn, D. N., et al. (2009). Phylogenetic analysis of seven WRKY genes across the palm subtribe Attaleinae (Arecaceae) identifies Syagrus as sister group of the coconut. PLoS One, 4(10), e7353. Published under Creative Commons License.*

Figure 2. Majority rule consensus of 12,500 trees after burn-in sampled from mixed model (partitioned) Bayesian analysis of seven combined WRKY loci sequences aligned across Areceaceae tribe Cocoseae subtribe Attaleinae with MrBayes. Numbers above branches are posterior probability scores, i.e., the proportion of tree within which that clade was resolved. Letter in red at nodes are ancestral area optimizations as determined by dispersal-vicariance analysis: A = South Africa, B = Madagascar, C = Amazonas (north, east, south), D = Chile, E = Eastern Brazil, F = Central Brazil, G = Andes, H = Central America, I = Mexico, J = West Indies, K = Southern Brazil, L = Northern South America, i.e., Caribbean coastal Venezuela and Colombia, French Guiana, Guyana, Surinam, M = Argentina-Paraguay-Uruguay, N = Colombia-Venezuela (*llanos* region), and O = western Amazonas. Ambiguous area optimizations at a node are separated by commas. doi:10.1371/journal.pone.0007353.g002

Figure 5.9 Majority-rule consensus of 12,500 trees after burn-in sampled from mixed model (partitioned) Bayesian analysis of seven combined WRKY loci sequences aligned across subtribe Attaleinae with Mr. Bayes. Detailed explanations are given beneath figure. *Reproduced from Meerow, A. W., Noblick, L., Borrone, J. W., Couvreur, T. L. P., Mauro-Herrera, M., Kuhn, D. N., et al. (2009). Phylogenetic analysis of seven WRKY genes across the palm subtribe Attaleinae (Arecaceae) identifies Syagrus as sister group of the coconut.* PLoS One, 4(10), e7353. *Published under Creative Commons License.*

The authors found further that the monophyly of the subtribe was unquestionable. The *Cocos* alliance resolved by *WRKY* genes consisted only of *Cocos nucifera*, *Syagrus*, and *Lytocaryum* (embedded in *Syagrus*), and it was positioned as sister to the *Attalea* alliance consisting of all the American genera. Meerow et al. (2009) then concluded that this study lent further support to the close relationship of *Lytocaryum* and *Syagrus*. Then they, went on to state that their data "presents the strongest evidence to date for a close phylogenetic relationship between *Cocos* and *Syagrus*." The authors have given some comments and dates on the age of differentiation and spread of the Cocoseae genera. We shall cover this in the next chapter.

Meerow et al. (2014) continued their studies from their earlier one (Meerow et al., 2009) on the phylogeny (and biogeography) of the entire tribe Cocoseae using 96 taxa plus two species each of *Roystonea* and *Reinhardtia* as outgroups. They employed six single-copy *WRKY* loci that had been isolated originally from *Cocos nucifera*. As experimental material, they used two coconut cultivars, Atlantic Tall and Niu Leka. The rest were *Acrocomia* (5 samples/3 species), *Astrocaryum* (11/6), *Aiphanes* (3/3), *Bactris* (11/10), *Desmoncus* (3/3), *Barcella* (1/1), *Elaeis* (3/2), *Allagoptera* (3/3), *Parajubaea* (2/2), *Attalea* (16/16), *Lytocaryum* (2/2), *Syagrus* (19/19), *Butia* (7/7), *Beccariophoenix* (1/2), *Jubaeopsis* (1/1), *Voaniola* (1/1), *Reinhardtia* (2/2), and *Roystonea* (2/2). They carried out concatenated maximum parsimony, ML, and Bayesian (BA) analyses, as well as three species–tree inference approaches. Their objectives consisted of testing the monophyly of all the genera and finding out if phylogenetic relationships with the Attaleinae would change with increased sampling of Elaeidinae and Bactridinae, as well as with the use of more distantly related outgroups than they had employed in their earlier (Meerow et al., 2009) work.

The authors concluded the following: (1) Subtribe Elaeidinae was sister to Bactridinae in all the analyses. (2) Within Attaleinae, *Lytocaryum*, previously nested in *Syagrus,* was positioned by MP and ML as sister to the latter with high support, and with BA, it continued to be embedded in *Syagrus*. (3) Both MP and ML resolved *Cocos* as sister to *Syagrus*, whereas BA positioned *Cocos* as sister to *Attalea*. (4) Bactridinae was composed of two sister clades, *Bactris + Desmoncus*, for which there was also morphological support; and the second, comprising *Acrocomia, Astrocaryum*, and *Aiphanes*. (5) The optimization of ancestral areas on the Bayesian supermatrix tree required 86 dispersals, 27 vicariance events (vicariance: A discontinuous biogeographical separation of an organism that previously inhabited a continuous range; King et al., 2006), and one extinction. (6) Although *Cocos* was still placed as sister to *Lytocaryum/Syagrus* subclade in the MP and ML supermatrix analysis (Fig. 5.10), the support was lower than in their earlier (2009) paper. (7) When the sister relationship of *Attalea* was tested as a constraint of the supermatrix with MP, it was found that only four base substitutions were enough to support the resolution. *Attalea* has a fibrous mesocarp as in *Cocos*, whereas *Syagrus* and *Lytocaryum* possess a fleshy mesocarp. In addition, the single-gene analysis of *WRKY* 6 gene resolved *Cocos* as sister to *Attalea*.

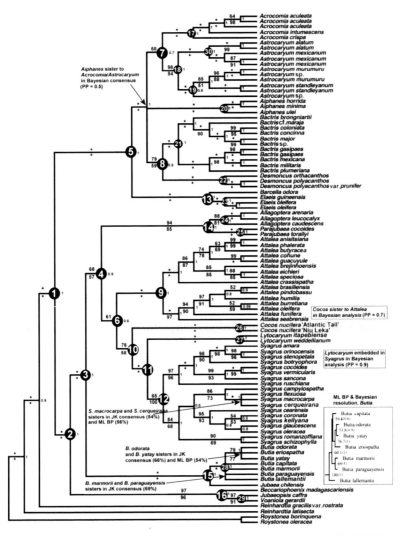

Fig. 1. Strict consensus tree of 12 equally parsimonious trees found by heuristic parsimony analysis of six combined WRKY loci sequences aligned across Arecaceae tribe Cocoseae, with *Roystonea* and *Reinhardtia* as outgroups (*Roystonea* is the functional outgroup). Numbers above branches are parsimony jackknife support percentages. Numbers below branches are ML bootstrap (BP) support (non-partitioned analysis). Numbers at nodes are posterior probabilities from the non-partitioned BEAST analysis. The numbered nodes are referenced in Table 2 (partitioned decay indices). The inset box shows the ML resolution of *Butia* with BP percentages. * = 100% support.

Figure 5.10 Strict consensus tree of equal parsimony found by heuristic parsimony (HP) analysis of seven combined *WRKY* loci sequences aligned across tribe Cocoseae. Detailed explanations given beneath the figure. *Reproduced with permission from Meerow, A. W., Noblick, L., Salas-Leiva, D. E., Sanchez, V., Francisco-Ortega, J., Jestrow, B., et al. (2014). Phylogeny and historic biogeography of cocosoid palms (Aracaceae, Arecoideae, Cocoseae) inferred from sequences of six WRKY gene family loci. Cladistics, 1–26.* http://dx.doi.org/10.1111/cls.12100.

Meerow et al. (2014) then concluded that "we are inclined to accept the sister relationship of *Cocos* and *Attalea* as the more likely scenario given the consensus of the two species–tree estimation approaches (Fig. 5.11) and two of the single-gene MP consensus topologies, as well as fruit morphology. Hydrochory (dispersal of seeds or spores by water), the dispersal mechanism of the coconut, is known in *Attalea* as well."

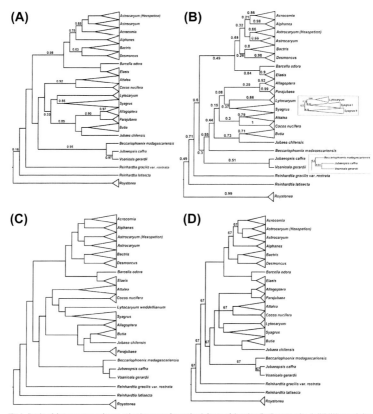

Fig. 3. Results of three gene-tree/species-tree estimation approaches on the phylogeny of the palm tribe Cocoseae using six WRKY transcription factor loci, and 50% majority rule consensus of supermatrix MP tree and two species trees. (a) Maximum clade credibility consensus from partitioned *BEAST analysis using BEAST Version 1.75. Numbers above or below branches are posterior probability (PP) scores. Only PP < 1.0 are shown. (b) Population tree (resolved by quartet solutions) from BUCKy analysis (Larget et al., 2010), with concordance factor scores above or below branches. Insets are resolution of two nodes from the primary CF tree (resolved by highest CF scores) that differed from the population tree. (c) Highest scoring ML species tree inferred from six WRKY loci consensus gene trees using STELLS. (d) Fifty per cent majority rule consensus tree of one of 12 equally parsimonious trees found by MP, the *BEAST species tree, and the BUCKy population tree. Only consensus indices < 100% are shown, above or below branches.

Figure 5.11 Results of three gene–tree/species–tree estimation approaches on the phylogeny of tribe Cocoseae using six *WRKY* transcription factor loci, and 50% majority rule consensus of supermatrix. Maximum-parsimony tree and two species of trees; details given later. *Reproduced with permission from Meerow, A. W., Noblick, L., Salas-Leiva, D. E., Sanchez, V., Francisco-Ortega, J., Jestrow, B., et al. (2014). Phylogeny and historic biogeography of cocosoid palms (Aracaceae, Arecoideae, Cocoseae) inferred from sequences of six WRKY gene family loci. Cladistics, 1–26. http://dx.doi.org/10.1111/cls.12100.*

Baker et al. (2009) carried out a comprehensive phylogenetic study of the entire palm family using all its genera (then, 192 in all). They compared the results obtained with both supertree and supermatrix methods. In addition, they used data obtained in all the earlier DNA sequence studies done by them and others till then (total, 15 studies), and also the morphological characters of all the genera. This procedure ensured including 22% of all the genera being represented by sequence from more than one species, while assuming generic monophyly. They ensured also that at least one molecular set was complete for all the palm genera and outgroups. They employed MRP (matrix representation with parsimony analysis) with bootstrap-weighted matrix elements, for analysis. This generated trees with maximum congruence with the supermatrix tree (Fig. 5.12). The authors further assessed cross-corroboration between supermatrix trees and the variant supertrees using strict consensus (Fig. 5.13). This yielded 160 clades (out of a maximum possible 204). Of these, 137 clades were present in both the trees.

Baker et al. (2009) study strongly supported the current phylogenetic classification of palms (cf. Dransfield et al., 2008). At the same time, in supertrees, some unorthodox relationships were recovered, such as Coryphoideae as sister to Ceroxyloideae and Calamoideae to Arecoideae. Also, the monophyly of three subfamilies including Arecoideae—to which the tribe Cocoseae belongs—and three tribes and four subtribes—but, no Cocoseae and Attaleinae—were questioned. In contrast, the supermatrix tree, the most congruent supertree, and most supertrees, was in substantial agreement with the palm classifications given in GP II (Dransfield et al., 2008), which included supporting the monophyly of all the subfamilies.

Incidentally, Baker et al. (2009) recovered sister relationship of *Cocos* with *Parajubaea* in both the supermatrix analysis and supertree analysis. Further, they found sister relationship between the tribe Cocoseae and Reinhardteae in all the four different analyses that the authors had carried out—super matrix, standard MRP with equal weights, standard MRP with bootstrap weights, and irreversible MRP with bootstrap weights (Fig. 5.14).

In a second study, the same group (Baker et al., 2011) studied the phylogenetic relationships among the arecoid palms, i.e., all the genera included in the subfamily Arecoideae (*sensu* GP II). This subfamily is the largest and most diverse of the five subfamilies of the palm family. It includes 13 tribes and 10 unplaced genera making a total of 108 genera and c. 1333 species. Out of these, they included 190 species/103 genera—total, c. 333 species/107 genera. They

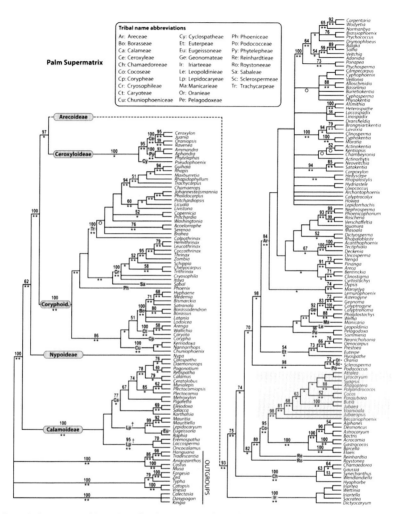

FIGURE 2. Strict consensus of the 46 080 trees from the supermatrix analysis of all palm genera (tree length = 15 173, CI = 0.41, RI = 0.62). Values above branches are BPs (>50%). Asterisks below branches indicate clades also recovered in supertrees; 1 asterisk denotes a clade recovered in the "most congruent supertree" (Fig. 2), 2 asterisks denote a clade recovered in all supertrees summarized in Figure 3 legend). An open circle indicates a clade absent from, but compatible with, the most congruent supertree, see Fig. 3 legend). An open circle indicates a clade absent from, but compatible with, the most congruent supertree. Palm subfamilies and tribes are indicated. Nonmonophyletic higher taxa are marked with a dagger symbol.

Figure 5.12 Strict consensus tree from the supermatrix analysis of all palm genera. Sub-tribe Attaleinae given in shade. *Reproduced with permission from Baker, W. J., Savolainen, V., Asmussen-Lange C. B., Chase, M. W., Dransfield, J., Forest, F., et al. (2009). Complete generic-level phylogenetic analyses of palms (Arecaceae) with comparisons of supertree and super-matrix approaches.* Systematic Biology. http://dx.doi.org/10.1093/syslro/sypo21.

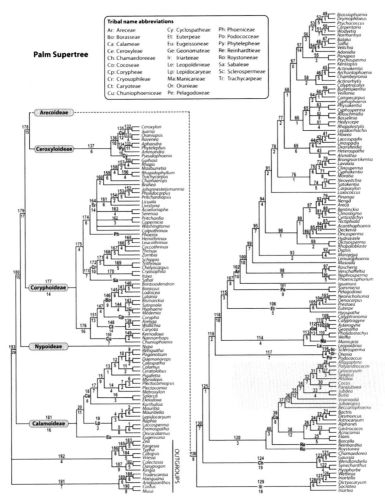

FIGURE 3. The "most congruent supertree." Strict consensus of 5000 supertrees from standard MRP analysis based on the expanded input tree set with matrix elements weighted in proportion to bootstrap values of corresponding input tree clades (maximum weight = 10, tree length = 11 830.667, CI = 0.80, RI = 0.95). Numbers above branches are clade numbers (see online Supplementary Appendix 1 for support values). Those below branches indicate the number of input trees that support a given clade (s). Palm subfamilies and tribes are indicated. Only Node 67 is not supported by any input tree.

Figure 5.13 Strict consensus of supertree from the standard matrix representation with parsimony (MRP) analysis. Subtribe Attaleinae in shaded portion. Details given beneath figure. *Reproduced with permission from Baker, W. J., Savolainen, V., Asmussen-Lange C. B., Chase, M. W., Dransfield, J., Forest, F., et al. (2009). Complete generic-level phylogenetic analyses of palms (Arecaceae) with comparisons of supertree and supermatrix approaches. Systematic Biology.* http://dx.doi.org/10.1093/syslro/sypo21.

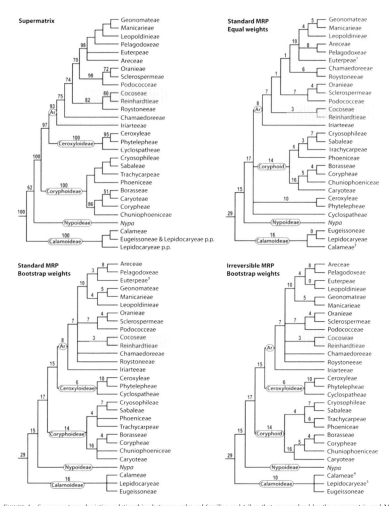

FIGURE 4. Summary trees depicting relationships between palm subfamilies and tribes that are resolved by the supermatrix and MRP analyses based on the expanded input tree set. The upper 2 trees show the relationships resolved by the supermatrix and equal weights standard MRP analyses. The lower 2 trees show relationships in common between all bootstrap-weighted standard MRP analyses and all bootstrap-weighted irreversible MRP analyses, respectively. The equal-weights irreversible MRP supertree is not shown here. Values above branches are BP's for the supermatrix tree and s for the supertrees. Palm subfamilies are indicated (Ar = Arecoideae). Notes: [1]Hyospathe resolves outside Euterpeae; [2]Calameae paraphyletic; [3]Hyospathe resolves outside Euterpeae in analyses with maximum weights = 1 and 2; [4]nonmonophyletic in analyses with maximum weights = 1 and 2; [5]nonmonophyletic in analyses with maximum weights = 1.

Figure 5.14 Summary trees depicting relationships among palm subfamilies and tribes that are resolved by supermatrix and matrix representation with parsimony (MRP) analyses. Tribe Cocoseae paired with Reinhardteae in all the trees. *Reproduced with permission from Baker, W. J., Savolainen, V., Asmussen-Lange, C. B., Chase, M. W., Dransfield, J., Forest, F., et al. (2009). Complete generic- level phylogenetic analyses of palms (Arecaceae) with comparisons of supertree and supermatrix approaches.* Systematic Biology. http://dx.doi.org/10.1093/syslro/sypo21.

analyzed the data using parsimony ratchet, maximum likelihood, and both likelihood and parsimony bootstrapping (Fig. 5.15). Then, they carried out the most densely sampled phylogenetic studies in this subfamily using DNA sequence data for the low-copy nuclear genes, *PRK* and *RPB2* (Fig. 5.16).

Their study strongly supported the current classification within the subfamily Arecoideae. However, they observed some incongruence between *PRK* and *RPB2* analyses, but the combined analysis helped to improve the resolution. The study supported three major higher level clades: (1) RRC clade (Roystoneae, Reinhardteae, and Cocoseae); (2) POS clade (Podococceae, Oranieae, and Sclerospermeae); and (3) the core arecoid clade (Arecaceae, Euterpeae, Genomateae, Leopoldinieae, Manicarieae, and Pelagodoxieae). *PRK* and *RPB2* yielded highly supported incongruent placements of Reinhardteae within the RRC clade, with the *PRK* placing

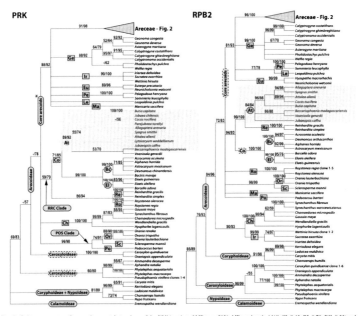

Fig. 1. Strict consensus trees from parsimony ratchet analyses of the *PRK* (number of MP trees, 3856; MP tree length, 1460; CI, 0·45; RI, 0·78; RC, 0·35) and *RPB2* (number of MP trees, 3742; MP tree length, 1562; CI, 0·52; RI, 0·79; RC, 0·41) data sets. Values above the branches are MP/ML bootstrap percentages. Groups recognized in the classification of Dransfield *et al.* (2005, 2008) and major clades mentioned in the text are indicated. The asterisk indicates that the core arecoid clade in this tree also includes tribe Iriarteeae. Labels with a dotted line indicate that the group is only resolved in part. Key to abbreviations: Ar, Areceae; Arc, Archontophoenicinae; Are, Arecinae; At, Attaleinae; Ba, Basseliniinae; Bc, Bactridinae; Ca, Carpoxylinae; Ch, Chamaedoreeae; Cl, Clinospermatinae; Co, Cocoseae; Dy, Dypsidinae; El, Elaeidinae; Eu, Euterpeae; Ge, Genomateae; Ir, Iriarteeae; Le, Leopoldinieae; Li, Linospadicinae; Ma, Manicarieae; On, Oncospermatinae; Or, Oranieae; Pe, Pelagodoxeae; Po, Podococceae; Pt, Ptychospermatinae; Re, Reinhardteae; Ro, Roystoneae; Sc, Sclerospermeae; Ve, Verschaffeltinae.

Figure 5.15 Strict consensus trees from parsimony ratchet analyses of the *PRK* and *RPB2* data sets. Subtribe Attaleinae in *shaded portion*. Details given beneath the figure. *Reproduced with permission from Baker, W.J., Norup, M. V., Clarkson, J. J., Couvreur, T. L. P., Dowe, J. L., Lewis, C. E., et al. (2011). Phylogenetic relationships among arecoid palms (Arecaceae: Arecoideae). Annals of Botany, 105, 1417–1436.*

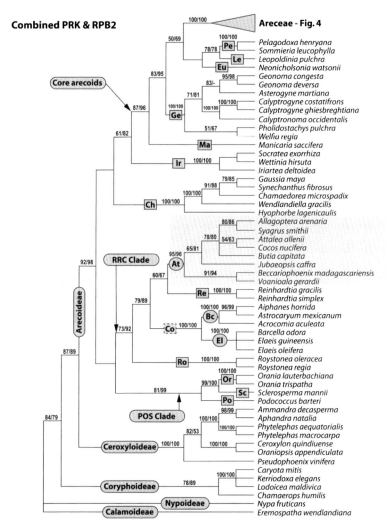

Combined PRK & RPB2

Areceae - Fig. 4

100/100
100/100
50/69
78/78 **Pe** — *Pelagodoxa henryana*
— *Sommieria leucophylla*
Le — *Leopoldinia pulchra*
83/95 **Eu** — *Neonicholsonia watsonii*
95/98 — *Geonoma congesta*
83/- — *Geonoma deversa*
71/81 — *Asterogyne martiana*
(Core arecoids) 100/100 100/100 — *Calyptrogyne costatifrons*
87/96 **Ge** — *Calyptrogyne ghiesbreghtiana*
— *Calyptronoma occidentalis*
61/82 51/67 — *Pholidostachys pulchra*
— *Welfia regia*
Ma — *Manicaria saccifera*
— *Socratea exorrhiza*
Ir 100/100 — *Wettinia hirsuta*
— *Iriartea deltoidea*
79/85 — *Gaussia maya*
91/88 — *Synechanthus fibrosus*
100/100 — *Chamaedorea microspadix*
Ch 100/100 — *Wendlandiella gracilis*
— *Hyophorbe lagenicaulis*
80/86 — *Allagoptera arenaria*
78/80 — *Syagrus smithii*
54/63 — *Attalea allenii*
65/81 — *Cocos nucifera*
95/96 — *Butia capitata*
92/98 **RRC Clade** **At** — *Jubaeopsis caffra*
60/67 91/94 — *Beccariophoenix madagascariensis*
— *Voanioala gerardii*
79/89 **Re** 100/100 — *Reinhardtia gracilis*
— *Reinhardtia simplex*
100/100 96/99 — *Aiphanes horrida*
Bc — *Astrocaryum mexicanum*
73/92 — *Acrocomia aculeata*
Co 100/100 100/100 — *Barcella odora*
El — *Elaeis guineensis*
— *Elaeis oleifera*
87/89 **Ro** 100/100 — *Roystonea oleracea*
— *Roystonea regia*
100/100 — *Orania lauterbachiana*
99/100 **Or** — *Orania trispatha*
81/99 **Sc** — *Sclerosperma mannii*
Po — *Podococcus barteri*
98/99 — *Ammandra decasperma*
POS Clade 100/100 — *Aphandra natalia*
82/53 100/100 — *Phytelephas aequatorialis*
— *Phytelephas macrocarpa*
84/79 (Ceroxyloideae) 100/100 100/100 — *Ceroxylon quindiuense*
— *Oraniopsis appendiculata*
— *Pseudophoenix vinifera*
100/100 — *Caryota mitis*
(Coryphoideae) 78/89 — *Kerriodoxa elegans*
— *Lodoicea maldivica*
Nypoideae — *Chamaerops humilis*
(Calamoideae) — *Nypa fruticans*
— *Eremospatha wendlandiana*

FIG. 3. Strict consensus trees from parsimony ratchet analyses of the combined analysis of *PRK* and *RPB2* (number of MP trees, 3662; MP tree length, 2855; CI, 0·48; RI, 0·74; RC, 0·36). See legend to Fig. 1 for further details and key to abbreviations.

Figure 5.16 Strict consensus trees from parsimony ratchet analyses of combined *PRK* and *RPB2* data sets. Subtribe Attaleinae given in *shade*. Details given beneath the figure. *Reproduced with permission from Baker, W.J., Norup, M. V., Clarkson, J. J., Couvreur, T. L. P., Dowe, J. L., Lewis, C. E., et al. (2011). Phylogenetic relationships among arecoid palms (Arecaceae: Arecoideae). Annals of Botany, 105, 1417–1436.*

it sister to Cocoseae, and the *RPB2* nesting it with Cocoseae as sister to subtribe Attaleinae (to which *Cocos* belongs).

In the study, the coconut *Cocos nucifera*, was seen as sister to *Parajubaea toralli* with *PRK* analysis, and as sister to *Attalea allessi* with both *RBP2* and combined *PRK* and *RBP2* analyses. The further higher-level relationship of this clade containing *Cocos* and *Parajubaea* (in *PRK* analysis) was with a clade of *Butia* and *Jubaea* in *PRK* analysis, that with *Cocos* and *Attalea* was with *Allagoptera* and *Syagrus* with *RPB2* analysis, and with *Allagoptera*, *Syagrus*, and *Butia* in the combined analysis.

Incidentally, the sister relationship of the two tribes Reinhardteae with Cocoseae is widely supported in most other studies as well (Asmussen et al., 2006; Baker et al., 2009; Hahn, 2002a).

Baker et al. (2011) concluded that within Cocoseae, *PRK* and *RPB2* and combined analyses supported the monophyly of the three subtribes and the sister relationship between Elaeidinae and Bactridinae, as obtained in several earlier studies (Asmussen et al., 2006; Baker et al., 2009; Eiserhardt, Swenning, Kissling, & Balslev, 2011; Gunn, 2004; Hahn, 2002a). Their findings also supported the earlier findings that the Neotropical genera of Attaleinae and pantropical *Cocos* formed a monophyletic group to the exclusion of the Madagascan (*Beccariophoenix* and *Voaniola*) and African (*Jubaeopsis*) genera of the tribe Cocoseae, as had been found earlier by several authors (Baker et al., 2009; Gunn, 2004; Eiserhardt et al., 2011; Meerow et al., 2009; Fig. 5.17).

In continuation of the study by Eiserhardt et al. (2011), Kissling et al. (2012) studied global palm species distribution, data, dated gene-level phylogeny, and a paleoreconstruction of biomes and climates. They found a strong imprint of phylogenetic clustering due to geographical isolation and in situ diversification in the Neotropics and on islands (e.g., Madagascar).

The summary trees depicting intertribe relationships (Fig. 5.17) of Baker et al. (2011) show some variation in the relationship of subtribe Attaleinae. It shows sister relationship with a clade containing the spiny subtribes of Bactridinae and Elaeidinae with *PRK* analysis, but with the tribe Reinhardteae with *RPB2* and combined *PRK* and *RPB2* analysis. The studies of Hahn (2002b), Asmussen et al. (2006), and the Baker et al. (2009) supertree show sister relationship of Attaleinae with the clades of Bactridinae and Eleidinae.

4. COMMENTARY

In the foregoing pages, we have summarized the work done so far on lineage differentiation in the palm family with emphasis on the coconut and the subtribe Attaleinae. Some other authors have given their perspectives also on this

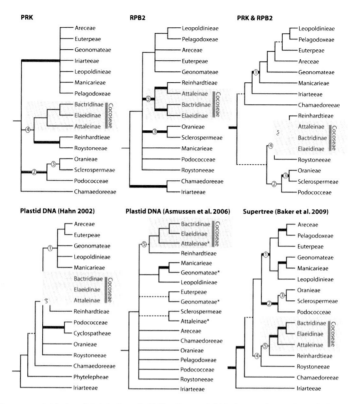

FIG. 5. Summary trees depicting inter-tribal relationships resolved in this study compared with the three most relevant previous studies (Hahn, 2002a; Asmussen et al., 2006; Baker et al., 2009). Note that tribe Pelagodoxeae was not sampled by Hahn (2002a). Plastid DNA regions sampled by Hahn (2002a) were atpB, rbcL, ndhF, trnQ–rps16 and trnD–trnT, and those sampled by Asmussen et al. (2006) were matK, rbcL, rps16 intron, trnL intron and trnL–F spacer. Baker et al (2009) combined 16 published data sets including those of Hahn and Asmussen et al. and existing data for PRK and RPB2 (e.g. Norup et al., 2006 Savolainen et al., 2006). Bold branches indicate relationships supported by bootstrap percentages ≥85 % (for both MP and ML, where available). For the super-tree of Baker et al. (2009), bold branches indicate relationships supported by five or more input trees (s ≥5). Remaining branches are supported by <85 BP (or for the supertree s <5) except for dotted branches that are not supported by >50 BP (or for the supertree s = 1). * indicates tribes that are not resolved as mono-phyletic. Sub-tribes of Areceae are not shown here. Key to clade annotations: 1, core arecoid clade (Areceae, Euterpeae, Geonomateae, Leopoldinieae, Manicarieae and Pelagodoxeae); 2, POS clade (Podococceae, Oranieae and Sclerospermeae); 3, Oranieae–Sclerospermeae clade; 4, RRC clade (Roystoneeae, Reinhardtieae and Cocoseae); and 5, Reinhardtieae–Cocoseae clade.

Figure 5.17 Summary trees depicting relationships resolved in this study compared with the three most relevant earlier studies (Asmussen et al., 2006; Baker et al., 2009; Hahn, 2002a). More details are beneath the figure. *Reproduced with permission from Baker, W.J., Norup, M. V., Clarkson, J. J., Couvreur, T. L. P., Dowe, J. L., Lewis, C. E., et al. (2011). Phylogenetic relationships among arecoid palms (Arecaceae: Arecoideae). Annals of Botany, 105, 1417–1436.*

aspect (e.g., Dransfield et al., 2008; Savolainen et al., 2006, etc.). All the authors had included in their studies representatives of all the genera, but not all the species in all of them. In addition, the coconut, as a widely cultivated species, does not appear to have been sampled or represented adequately in any of these studies. A percent sampling of all the species may not be possible, mainly because the taxonomy of all the genera has not been studied, and further, live collections of all the species do not appear available anywhere.

Several of the palm species have been used by indigenous people for various purposes in South America, albeit at a moderate level, and to that extent, they might have undergone at least a minimal level of ennoblement as well. Hence, even if the same species has been included by more than one study, their actual samples could have been genetically different.

Though all these studies have given broadly comparable results at the higher levels of the taxa, variations in topologies (and divergence times) at lower levels are noticeable. This difference seems to become wider, as we move down the taxonomic hierarchy. At the same time, the monophyly of the tribes, as delimited now, and practically of all the taxa up to the genus level, have been confirmed in all the studies.

The earlier suggestion (Hahn, 2002b; Uhl & Dransfield, 1987) that the distribution pattern of the tribe was reflective of Gondwanan breakup has been disproved. All the authors have indicated that the earliest divergences within the tribe Cocoseae initially took place in South America.

There is, however, no agreement on two points: (1) The initial mode of dispersal of subtribe Attaleinae from South America; and (2) the extant genus having the closest relationship with the genus *Cocos*.

There are different opinions also about the dispersal of subtribe Attaleinae and Cocoseae's relationship with other tribes. Although we shall discuss the dispersal aspects here, we shall consider the second point—the genus most closely related to the genus *Cocos*—in a later chapter on Origin.

There are two opinions with regard to the initial dispersal of the subtribe Attaleinae from South America—via the Antarctic land bridge from South America into Eurasia or westward into the Pacific Ocean from South America.

Of 18 genera in the tribe Cocoseae, two occur only in Madagascar [*Beccariophoenix* (two species) and *Voaniola* (one species)]; a third one [*Jubaeopsis* (one species)] is endemic to eastern South Africa, and a fourth genus, *Elaeis* (two species), occurs in both South America and West Africa (with one species each). *E. guineensis* (oil palm), which is, incidentally, the most productive of all the oil-yielding plants, is native to tropical West Africa. It occurs there in wild and part-domesticated states (Zeven, 1967). Presently, 90% of oil-palm cultivation is carried out in Southeast Asia, almost entirely in Indonesia and Malaysia. The second species, *E. oleifera,* is native to central and northern South America. *Cocos*, the coconut, is pantropical in cultivation. All the remaining 13 genera of the tribe Cocoseae occur in only Central and South America, and the Caribbean Islands. There are also certain other intriguing aspects also about the biogeography of *Attalea* palms. We shall consider them in the next chapter.

CHAPTER 6

Biogeography

1. INTRODUCTION

Biogeography is "the study of the distribution of organisms over the earth and of the principles that govern these[sic] distribution" (King, Stansfield, & Mulligan, 2006).

The coconut has a pantropical distribution. It is basically a strand plant. It extends also to lowland subtropics in the areas that receive 1000–1500 mm of distributed precipitation annually. Interestingly, and broadly stated, the distribution range of the coconut matches that of the tribe Cocoseae, and subtribe Attaleinae to which it belongs (Fig. 6.1). The natural habitats of these three entities, however, differ according to their respective ecological preferences.

Presently, Almost the entire coconuts that we find throughout the world is cultivated. However, there are still some isolated pockets and several islands where it occurs in natural strands in the lowland coastlands and in much of the thousands of islands of the Indian and Pacific Oceans. This is evident from paleobotanical and paleopalynological evidences. For instance, naturally occurring forms of coconut were present as a component of the natural vegetation in the Cook Islands—and possibly even as far as the Society Islands—from the western Pacific Ocean, long before the present-day humans reached there (Bellwood, 2013; Kirch, 2000).

Dransfield et al. (2008; GP II) have given meta-accounts of the distribution and ecology of the family Arecaceae up to its tribes, genera, and species. Palms broadly occur throughout the tropics and subtropics of the world. Tropical rain forests and tropical islands of the Old World constitute their natural habitats. At the same time, some palms also occur in seasonal and semiarid and moderately cool habitats (e.g., cerrado of central Brazil), and rarely, in desert locations where groundwater is available, and at high latitudes; e.g., at 44° N in Mediterranean France, 44° S in the Chatham Islands, east of New Zealand (Dransfield et al., 2008). Some authors have observed that the family's latitudinal extremes are largely limited by those of subfamily Coryphoideae and tribe Cocoseae.

The Arecaceae now include 183 genera and c. 2400 species (2588 species in 188 genera per Palmweb, accessed February 25, 2016). Malesia (Sumatra,

The Coconut
ISBN 978-0-12-809778-6
http://dx.doi.org/10.1016/B978-0-12-809778-6.00006-1

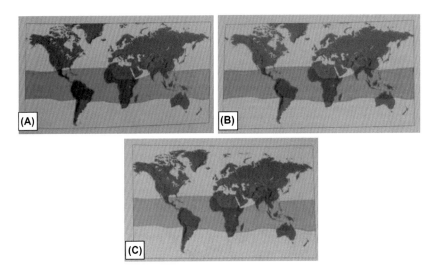

Figure 6.1 Present distribution of (A) Tribe Cocoseae. (B) Subtribe Attaleinae. (C) Genus *Cocos*.

Malay Peninsula, Philippines to New Guinea, and the Solomon Islands) has the largest concentration of palms (50 genera, c. 992 species). When tropical Asia is included within Malesia, the diversity expands to 57 genera and c. 1200 species (Dransfield et al., 2008). They attribute this to prolific speciation in island archipelagoes and the admixing of Northern and Southern Hemisphere floras in this region due their coming together at the eastern and western ends of Malesia during the latter half of the Tertiary (Tertiary, 66–5 Mya). A fair level of climatic stability during this period facilitated this process. In palm diversity, Malesia is followed by the Neotropics (65 genera, c. 130 species). Numerous tribes of the subfamily Arecoideae dominate here, including the tribe Cocoseae (Dransfield et al., 2008). Henderson, Galeano, and Bernal (1995) had pointed out that just three Arecoid genera, *Chameodorea, Geonoma,* and *Bactris,* account for one-third of the palm diversity in the Neotropics. The western Indian Ocean islands (including Madagascar) is another region rich in palm diversity (25 genera, 192 species) (Dransfield et al., 2008).

2. AGE AND LINEAGE DIFFERENTIATION OF PALMS

Traditionally, the phylogenetic ages of different groups and their lineages are estimated on the basis of fossils. Palms possess a fairly rich collection of fossils of roots, stems, leaves, flowers, and pollen. However, most of them have only been of limited use in estimating the age for several reasons, as we have noted before. Further, correlating the fossil taxa with the extant ones has been also problematic, more as a rule than as an exception (Harley, 2006).

More recently, molecular biologists have employed molecular clock methods to estimate the time of origin of plants and animals at even the lowest taxonomic level of species. Molecular clocks are known also as deoxyribonucleic acid (DNA) clocks or protein clocks. DNA clock hypothesis is the "postulation that when, averaged across the entire genome of a species, the rate of nucleotide substitutions remains constant. Hence, the degree of divergence in nucleotide sequences between two species can be estimated to use their diverge node" (King et al., 2006). Likewise, protein clock hypothesis is "the postulation that amino acid substitutions take place at a constant rate for a given family of proteins (e.g., cytochrome, hemoglobins), and hence, the degree of divergence between two species in the amino acid sequences in the protein in question can be used to estimate the length of time that has elapsed since their divergence from a common ancestor" (King et al., 2006).

The most widely cited instance of estimating the age of a plant taxon successfully using fossils has been the observation of Stebbins (1981) that the herbivores and grasses co-evolved during the lower to middle Miocene (Miocene: 57–36 Mya) strata in the South American grasslands. Subsequent studies using molecular clocks have largely confirmed this date (cf. Doyle & Gaut, 2000; Gaut, 2002; Soltis & Soltis, 2000, 2003). In a review and critique of this method, Bromham and Penny (2003) observed that their reliability depends on the accuracy with which genetic distance is estimated and the appropriateness of the calibration methods. In the absence of alternate and more reliable methods, this method does provide useful estimates about the relative phylogenetic diversifications timings at various levels of taxonomic entities.

The earliest known monocot fossil is of Upper Barremian or Lower Aptian age, c. 110–120 Mya (millions of years ago) (Friis, Penderson, & Crane, 2004). Nevertheless, now Barremian Age is estimated as 126–131 Mya. For the diversification of palms, the few studies have given slightly differing results—100 Mya (Bremer, 2000), 91–99 Mya (Wikstroem, Savolainen, & Chase, 2001), and 120 Mya (Janssen & Bremer, 2004).

Alluding to this, Dransfield et al. (2008) observed that the previous estimates would imply that the palms appear to have originated only after the initial breakup of Gondwana, which was underway by 130 Mya

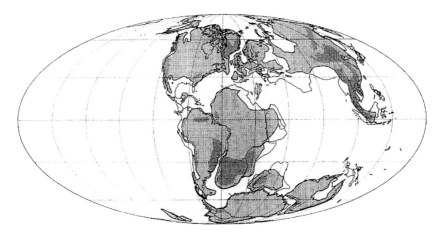

Figure 6.2 World paleocoastline map 130 Mya (Barremian–Hauterivian age). *Reproduced with permission from Smith, A. G., Smith, D. G., & Funnell, B. M. (1994).* Atlas of Mesozoic and Cenozoic coastlines *(99 pp.). Cambridge: Cambridge University Press.*

(see Fig. 6.2). By this time, sea gaps had existed between most of the landmasses (cf. Smith, Smith, & Funnell, 1994). Consequently, "oceanic dispersal has undoubtedly played an important role in establishing palms as component of tropical ecosystems" in the world (Dransfield et al., 2008).

The earliest unequivocal palm fossils, as we have noted before, are from the late Mid-Cretaceous to early Upper Cretaceous (Harley, 2006). They consisted of costapalmate leaves of three *Sabalites* species from Late Coniacian–Early Santonian (86–76 Mya) of South Carolina and from Santonian (82–76 Mya) of New Jersey [both in northeastern USA (Berry, 1914)]. The first record of palm fruits was from the Upper Cretaceous of Brazil (*Palmocarpon luisii*; Maury, 1930).

3. EXPERIMENTAL STUDIES

There have been only a handful of studies in this area in palms in general and in the Cocoseae in particular (about 15 nos.)

Hahn (2002a, 2002b) suggested that the Arecoid line was Gondwanan in origin with several independent dispersals into the Indo-Pacific region (Indian Ocean islands including Madagascar, plus Asia and the South Pacific). Cocoseae elements were possibly present on more than one fragment of the land, with the result that primary diversification of this group might have coincided with continental breakup.

This implies that phylogenetic branching patterns should reflect the timing and sequence of tectonic events. To recall, out of 18 genera in the tribe Cocoseae, two genera (*Beccariophoenix* and *Voaniola*) occur in only Madagascar, one genus is endemic in eastern South Africa (*Jubaeopsis*), and one genus (*Elaeis*) has one species each in tropical West Africa and the Neotropics. The remaining 14 genera occur in only the Neotropics. Recent phylogenetic studies have shown that Gondwanan breakup did not appear to have had any relevance in the distribution of the tribe Cocoseae (Baker & Couvreur, 2013a, 2013b; Dransfield et al., 2008). See also Table 6.1.

At the same time, Hahn had indicated that the dispersal of the genus *Cocos* was a non-Gondwanan element. The phylogenetic resolution of the genus obtained by him suggested a westward dispersal from South America. The discovery of an Attaleinae fossil endocarp from Easter Island (Dransfield, Flenley, King, Harkness, & Rapu, 1984) led Hahn to suggest a complicated story of repeated dispersals of Butiinae and Attaleinae into the region followed by extinction. The author did not elaborate the proposal further.

The first experimental investigation into this aspect was of Gunn (2004). She used three *Cocos nucifera* varieties (two tall and one dwarf) in her molecular analysis of the Cocos Alliance (*sensu* Moore) genera. To estimate divergence timings, Gunn used two fossils—Boulder Hill, New Zealand (42 Mya) and *Cocos intertrappeansis*, India (50 Mya).

Based on her analyses, Gunn (2004) then proposed that Cocoeae (sensu Moore, 1973) might have originated in the Paleocene, diversified and spread eastward to Africa, Madagascar, and India, and southward to Australasia and New Zealand through the West Antarctic corridor (which was extant till the Oligocene). She felt that the putative progenitors of the coconut were from South America. Cocoeae then migrated to India from Madagascar after the divergence of the basal clade via long distance dispersal, because the fossil endocarps found in India were of lower Eocene age, and by this time, India had already separated from Madagascar.

She estimated then that Cocoeae diverged 53.26–61.09 Mya and *Cocos nucifera* 22.20–26.84 Mya (Table 6.1).

The second experimental study was done by Meerow et al. (2009). They conducted phylogenetic analysis of the subtribe Attaleinae, tribe Cocoseae, with 72 samples and using seven *WRKY* genes. The authors included six coconut cultivars, Atlantic Tall, Green Malayan, Panama Tall, Red Malayan, Red Spicata, and Nieu Leka. They conducted molecular dating using BEAST 1.4.8 and used as standard an Attaleinae fossil fruit from northern Colombia dated to c. 60 Mya. The authors had designated this fossil as "cf.

Table 6.1 Estimated divergence times of clades in tribe Cocoseae

Chronograms/Mya	1	2	3	4
Outgroups: *Reinhardtia and* *Roystonea*	–	56.12(N1)	96.61(N1)	75.31(N1)
Cocoseae	–	53.26(N2)	61.09(N2)	60.71(N2)
Spiny/nonspiny divergence	60.45	50(F2)	50(F2)	50(F2)
Nonspiny clade	51.02(N25)	43.07(N4)	42.88(N5)	42.93(N)
Spiny clade	42(F1)	46.29(N43)	46.28(N44)	46.28(N45)
Malagasy clade	42.06(N26)	39.64(N5)	39.42(N6)	39.48(N7)
Jubaeopsis caffra	35.72(N50)	37.84(N6)	37.62(N7)	37.68(N8)
Cocos alliance	37.64(N27)	34.16(N19)	33.96(N20)	34.01(N21)
Attalea alliance	34.87(N51)	37.84(N6)	37.62(N7)	37.68(N8)
Parajubaea torallyi	27.87(N33)	29.62(N21)	29.44(N22)	29.48(N23)
Cocos nucifera	22.20(N35)	26.84(N22)	26.67(N23)	26.71(N24)
Elaeis guineensis + *E. oleifera*	–	42(F1)	42(F1)	42(F1)

Comparisons of estimated times of divergences of clades from chronograms (not shown) 1 and 2, 3, and 4 using different fossil calibration points.
N, node; *F1*, Boulder Hill, New Zealand (42 Mya), *F2*, *Cocos intertrappeansis*, India (50 Mya), Cocoseae fossils for setting minimum age constraints. Columns 1, 2, 3, and 4 represent the estimated times of divergence of the clades. Cocos alliance = [*Cocos nucifera* + (*Allagoptera* + *Polyandrococos*) + (*Butia* + *Jubaea*) + (*Syagrus amara* + *S. smithii*) + *Parajubaea*]. Attalea alliance = [*Attalea* + (*Syagrus romanzoffiana* + *Lytocaryum weddellianum*)]. Outgroups: *Reinhardtia gracilis* and *Roystonea regia*.
Reproduced with permission from Gunn, B. F. (2004). Phylogeny of the Cocoeae (Arecaceae) with emphasis on *Cocos nucifera. Annals of the Missouri Botanical Garden, 91,* 505–525.

Cocos." Meerow et al. (2009) "conservatively" placed the fossil at the stem node of Attaleinae.

Meerow et al. showed that *Cocos* strongly shared a sister relationship with *Syagrus.* The genera *Cocos* and *Syagrus* diverged c. 35 Mya (Table 6.2). The crown node of the "modern" *Cocos* was estimated at 11 Mya. The authors then went on to conclude that "the biogeographic ancestry of the coconut, regardless of its subsequent ethnobotanical history is(was) firmly rooted in South America." Incidental to this are the findings of the authors about the close relationships found among *Allagoptera, Parajubaea,* and *Polyandrococcus.*

The authors noted further that direct land connection between Africa and South America was severed in Late Cretaceous (86–70 Mya), or the latest by Early Paleocene (65–62 Mya), possibly via the Walvis Ridge (Namibia, southern Africa)/Rio Grande Rise (deep south of Brazil) and Sierra Leone Ridges (present-day Sierra Leone, West Africa). Incidentally,

Table 6.2 Estimated divergence dates for selected dates within tribe Cocoseae, subtribe Attaleinae

Clade	Mean (MyBP)	95% HPD[a] lower bound	95% HPD upper bound
Stem node Attaleinae (calibration)	60.0	57.0	62.8
TMRCA[b] Attaleinae	43.7	27.2	50.3
TMRCA African Attaleinae	28.5	16.6	41.6
TMRCA American Attaleinae	38.4	23.9	44.4
TMRCA Allagoptera/Polyandrococos	20.4	9.9	23.3
TMRCA *Attalea*	13.0	8.6	20.5
TMRCA *Butia*	8.1	4.5	10.9
TMRCA *Butia/Jubaea*	14.5	8.2	20.6
TMRCA *Cocos/Syagrus/Lytocaryum*	34.9	20.7	39.5
TMRCA *Syagrus* (inc. *Lytocaryum*)	27.0	15.4	30.5
TMRCA *Parajubea/Butia/Jubaea/Allagoptera*	31.6	17.4	34.2
TMRCA Outgroup	50.7	41.1	63.3

Doi:10.1371/journal.pone.0007353.00s.
[a]HPD = Highest Posterior Density.
[b]TMRCA=Time of most recent common ancestor.
Reproduced from Meerow, A. W., Noblick, L., Borrone, J. W., Couvreur, T. L. P., Mauro-Herrera, M., Kuhn, D. N., et al. (2009). Phylogenetic analysis of seven *WRKY* genes across the palm subtribe Attaleinae (Aracaceae) identifies *Syagrus* as sister group of the coconut. *PLoS One*, 4(10), e7353. Creative Commons License.

according to Smith et al. (1994), the last connection between South America and Africa was at Sierra Leone Ridge and this was broken as early as during 105−95 Mya.

Meerow et al. observed further that the palms diversified greatly in northern South America. The stem age of the subtribe Attaleinae fell in Late Paleocene (66–57 Mya). They found it difficult to interpret the paleohistorical connection of the subtribe Attaleinae's three African genera with the other Attaleinae genera of South America. In contrast, the establishment of two Neotropical clades appeared congruent with the terminal Eocene cooling event that had happened at 38.4 Mya, they pointed out. In support of the previous contention, they stated that all the ancestral Neotropical Attaleinae were restricted to eastern Brazil. Meerow et al. (2009) had followed up this study using additional material and newer programs. Their findings are given later in the chapter.

Couvreur, Forest, and Baker (2011) had shown that the extant lineages of palms diversified initially in Laurasia (the present North America, Europe, and Asia before the breakup) toward the end of the Early Cretaceous—around the Albian−Cenomanian boundary, c. 100 Mya. Incidentally, this corresponds with the age of the earliest palm fossils. They had been reported from the Cretaceous of Europe and North America (Harley, 2006). The authors observed that there was mounting evidence that dispersal, rather than Gondwanan vicariance, was the major mechanism behind the pantropical distribution of certain tropical rain-forest plant families (including the family Arecaceae).

In a comprehensive study, Baker and Couvreur (2013a, 2013b) attempted to obtain evolutionary explanations for the present geographical distribution of various palm lineages. For this, they used the chronogram (see Fig. 6.3) of Couvreur et al. (2011). This chronogram had itself been based on the complete genus-level supertree of palms of Baker et al. (2009). The authors analyzed the data by employing a "released clock with uncorrelated rates" using the BEAST 1.5.3 program. To estimate the ancestral areas, the authors employed the maximum likelihood (ML) dispersal−extinction−cladogenesis method. Genera were assigned to eight geographical regions based on published information on genera distribution in Dransfield et al. (2008). These regions were **A**: South America; **B**: North and Central America; **C**: Africa; **D**: Indian Ocean; **E**: India; **F**: Rest of Eurasia, and **G**: Pacific Ocean (Table 6.3). They assigned *Cocos* to S. America because of it being nested among South American genera. Incidentally, this was in agreement with the findings of Meerow et al. (2009) also.

The authors then made the following observations. Only the information relevant to the coconut is presented here. (1) Around 100 Mya, before the end of the Cretaceous, the crown node of divergence of palms of all the five subfamilies occurred in Laurasia (Eurasia plus North America). (2) From Laurasia, palms dispersed on 35 occasions into all the major tropical areas (Table 6.4). (3) Beginning 50 Mya, six dispersals possibly took place on different occasions into the Pacific region east of the Wallace Line. (4) Then, a complex pattern of westward expansion into the Pacific Ocean and extinction in Eurasia took place, followed by eastward expansion into Eurasia and extinction in Pacific Ocean. (5) Many of the migrations across the Wallace Line inferred here predated the Miocene (23−5 Mya). (6) Prior to the clade's crown-node divergence (78.29 Mya, stem; 73.63 Mya, crown), the stem lineage of the subfamily Arecoideae had expanded into South America. (7) Many early divergences took place in South America: Rhienhardteae and Cocoseae at 59 Mya; range 55–64 Mya; tribe Cocoseae, at 59.43 Mya, stem/55.77 Mya, crown. (8) Within Cocoseae, the stem-node divergence also took place in S. America into subtribes Attaleinae

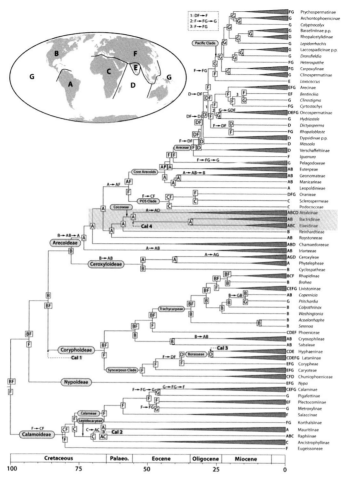

Figure 1 Spatial and temporal dimensions of palm evolution. Chronogram for the palm family, summarized where possible to the tribal and subtribal level (see also Table 1, Appendices S2 & S3, and tree file at the Dryad data repository, doi:10.5061/dryad.vb25b35j). For taxonomy see Dransfield *et al.* (2008). Calibration points are marked in grey boxes – Cal 1: *Sabalites carolinensis* (Berry, 1914), Cal 2: *Mauritiidites* (Schrank, 1994), Cal 3: *Hyphaene kapelmanii* (Pan *et al.*, 2006), Cal 4: fossil Attaleinae (Gomez-Navarro *et al.*, 2009). The most likely ancestral areas determined with LAGRANGE under model M₁ (Couvreur *et al.*, 2011a) are given in boxes at each node. For alternative reconstructions within two log-likelihood units see raw LAGRANGE output at Dryad data repository, doi:10.5061/dryad. vb25b35j. The inset map illustrates geographical areas defined for this study – A: South America, B: North America (including Central America and the Caribbean), C: Africa (including Arabia), D: Indian Ocean Islands (including Madagascar) E: India (including Sri Lanka), F: Eurasia (to Wallace's Line, including Andaman and Nicobar islands) and G: the Pacific (including areas east of Wallace's Line and Australia). Inferred range expansion events are marked where appropriate on internal branches. Due to the lack of phylogenetic resolution and ancestral area reconstructions within genera, range expansion inferences are not made on terminal (i.e. genus-level) branches.

Figure 6.3 Spatial and temporal history of dispersal events in the palms. *Reproduced with permission from Baker, W. J., & Couvreur, T. L. P. (2013a). Global biogeography and diversification of palms sheds light on the evolution of tropical lineages I. Historical biogeography.* Journal of Biogeography, 40, *274–285.*

Table 6.3 Estimated divergence times up to tribe levels in palm family stem and crown node estimates with 95%, Highest Posterior Density (HPD) values

	Stem	HPD upper	HPD lower	Crown	HPD upper	HPD lower
Family Arecaceae	117.9	113.8	120.0	100.0	92.0	108.8
Subfamily Arecoideae	78.3	70.7	85.3	73.6	66.2	81.4
Tribe Cocoseae	59.4	55.7	64.0	55.8	54.8	57.7
Subtribe Attaleinae	55.8	54.8	57.7	36.2	23.3	49.8
Subtribe Bactridinae	35.0	20.0	50.6	22.3	12.0	34.2
Subtribe Elaeidinae	35.0	20.0	50.6	15.6	3.9	30.3

Reproduced with permission from Baker, W. J., & Couvreur T. L. P. (2013a). Global biogeography and diversification of palms sheds light on the evolution of tropical lineages I. Historical biogeography. *Journal of Biogeography, 40*, 274–285.

Table 6.4 Estimated divergence times of selected clades within tribe Cocoseae subtribe Attaleinae

Taxon	Stem age (Ma)	95% HPD interval (Ma)	Crown age (Ma)	95% HPD interval (Ma)
Acrocomia	33.04	26.42–40.99	16.62	9.19–24.43
Aiphanes	34.80	28.71–42.38	14.71	7.84–22.06
(*Allagoptera, Parajubaea*)	28.68	22.27–35.57	22.24	15.52–29.23
Allagoptera	22.24	15.52–29.23	16.18	9.74–22.53
Astrocaryum	33.04	26.42–40.99	28.09	20.35–36.43
Attalea	23.61	17.35–30.51	12.82	8.68–17.57
Attaleinae	63.81	62.00–66.97	37.84	28.79–47.08
Afro-Malagasy Attaleinae	37.84	28.79–47.08	30.89	19.23–42.27
Bactridinae	56.21	48.10–64.51	36.06	30.00–42.53
Bactris	33.30	27.01–40.35	20.68	15.68–26.69
(*Butia, Jubaea*)	31.40	23.82–38.59	14.71	8.87–21.39
Butia	14.71	8.87–21.39	7.5	4.61–10.89
Cocos	23.61	17.35–30.51	3.92	1.34–6.78
Cocoseae	66.26	62.00–70.60	63.81	62.00–68.22
Desmoncus	33.30	27.01–40.35	14.19	8.12–20.59
Elaeidinae	56.21	48.10–64.51	23.74	20.00–30.65
Elaeis	23.74	20.00–30.65	7.38	2.86–12.69
Jubaea	14.71	8.87–21.39	n/a	n/a
Lytocaryum	17.09	12.80–22.05	6.47	2.94–10.98
Parajubaea	22.24	15.52–29.23	3.83	1.20–6.86
Syagrus (*Lytocaryum* embedded)	27.60	21.23–34.32	19.98	14.99–24.95

Reproduced with permission from Meerow, A. W., Noblick, L., Salas-Leiva, D. E., Sanchez, V., Francisco-Ortega, J., Jestrow, B., et al. (2014). Phylogeny and historic biogeography of cocosoid palms (Aracaceae, Arecoideae, Cocoseae) inferred from sequences of six *WRKY* gene family loci. *Cladistics,* 1–26. http://dx.doi.org/doi:10.1111/cls.12100.

(56 Mya), Bactridinae (35 Mya), and Elaeidinae (23–50 Mya). (9) The sub-tribe Attaleinae expanded into the Indian Ocean islands by the crown-node age of 36 Mya (range, 23–50 Mya), and into Africa after 29 Mya (range, 23–50 Mya). (10) Constant diversification continued till the end-Oligocene. However, some speciation bursts did also occur in recent times.

(11) Pan, Jacobs, Dransfield, and Baker (2006) and others had pro-posed that depauperization of Africa could be linked to the deteriorat-ing climate in the continent since the Miocene. (12) The diverse palm floras in many islands—Madagascar, Hawaii, New Caledonia—were attributed to increased diversification rates. (13) Above the genus level, 35 dispersal events between regions were inferred to have happened. The Indian and Pacific Oceans were the most frequent sinks of palm diversity (eight lineages each). The remaining dispersal events consisted of four events each to South America and Africa, and one into India. (14) Although Laurasia (Eurasia plus North America) was the most likely ancestral area for the crown node of palms, subsequent back dispersals into these regions appear to have taken place and, importantly, with seven dispersals into Eurasia and three into N. America. (15) Eurasia was the most important source of dispersal lineages (12 nos.); then, it was S. America (7 nos.), followed by Pacific Ocean (6 nos.), then Indian Ocean (5 nos.), and lastly N. America (4 nos.). Africa and India contributed only a little (one and zero, respectively). (16) The high palm diversity of S. America (c. 730 species) was almost entirely derived from just four major dispersal events into the continent—three from N. and C. America and one from Africa. (17) The high species diversity in S. America indi-cates that in situ diversification, not immigration, must be the most significant explanation. The large majority of S. American diversity could be traced back to a single Cretaceous dispersal event in the stem lineage of Arecoideae, with the remaining three dispersals giving only moderate species numbers. (18) Numerous early divergences in Arecoi-deae took place in S. America, which gave rise to many species-rich lineages, such as Cocoseae (300 spp.), Genomatiae (c. 80 spp.), Iriarteae (c. 30 spp.), and Tnerpeae (c. 30 spp.). (19) S. America was the source of many lineages dispersing to N. America, Eurasia, the Indian Ocean, and the Pacific Ocean.

The authors could not explain the long-distance dispersals of tropical groups. However, a dispersal route for megathermal taxa via Antarctica was possible during the climatic option of the Paleocene and Miocene with lineages diverging when once the climate barrier was reinstated. They inferred that palm lineages dispersals from South America to Indian Ocean and Pacific Ocean occurred between 55 and 40 Mya, which might be explained by migration through Antarctica during the Early Eocene thermal maximum (Figs. 6.2 and 6.4). (20) Only one dispersal into India appeared to have taken place—from the Indian Ocean in the Miocene. This suggests that the present-day Indian palm flora has been the result of recent colonization, rather than ancient autochthonous lineages. However, the presence of Cenozoic palm fossils of uncertain affinity indicate that ancient dispersals did take place into India. The rapid latitudinal change in Indian position during the Early Cretaceous until collision with Eurasia c. 35 Mya (Ali & Aitchison, 2008) is also assumed to have resulted in numerous lineages (Morley, 2000). (21) Dispersals between Eurasia, the Pacific, and Indian Oceans were likely facilitated by the geological events associated with the Miocene collision of the Australian plate with Southeast Asia (Hall, 2002). The resultant Malaysian archipelago would have provided adequate environmental milieu to permit new lineages to develop. (22) The authors have inferred six dispersals across the Wallace Line to the east, and four westward dispersals, happening mostly in the Eocene. This supports the proposition of Richardson, Costion, and Muellner (2012) that eastward migrations across the Wallace Line were more common than westward ones.

Based on the above inferences, Baker and Couvreur (2013b) opined that palm evolution was the result of a mixed model of diversification with an overall steady accumulation of major lineages punctuated by steady shifts in diversification. All of them, except one, took place within the last 30 My, with eight within the last 20 My. Thus, most shifts could be attributed to recent events. They proposed also an important role for Malesia in palm diversification-rate shifts on the basis of a high number of genera identified to have undergone a rate increase and showing high rates of diversification.

The study by Baker and Couvreur (2013b) shows that the earliest divergences in the Cocoseae took place initially in South America (see Fig. 6.4) during the Paleocene (66−57 Mya) and Early Eocene (Eocene, 55–57 Mya). The authors have, however, cautioned that their ancestral area reconstruction (AAR) was unable to explain the complete paleodistribution of the tribe Cocoseae. For instance, fossils belonging to Cocoseae recovered from New Zealand have been attributed to Eocene (57−36 Mya). In addition,

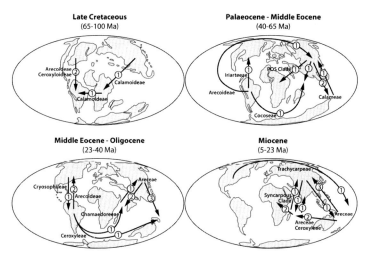

Figure 2 Summary of the dispersal history of palms. Four time frames were chosen to represent the history of the major dispersal events in palms. Arrows represent inferred dispersal events (*not* specific dispersal routes) above the genus level following Fig. 1. Circles on arrows indicate the number of dispersal events that took place in the same direction. Subfamily, tribe or major clade names are indicated on certain arrows (Dransfield *et al.*, 2008). Base maps are derived from Buerki *et al.* (2011).

Figure 6.4 Summary of the dispersal history of palms in four time frames representing the major dispersal events. *Reproduced with permission from Baker, W. J., & Couvreur, T. L. P. (2013b). Global biogeography and diversification of palm sheds light on the evolution of tropical lineage. II. Diversification history and origin of regional lineages. Journal of Biogeography, 40, 286–298. Base maps are derived from Buerki et al. (2011).*

Dransfield et al. (1984) had recorded a cocosoid palm dated to 800Ya from Easter Island. The authors had found further that the first expansion of Arecoideae from South America to Eurasia occurred during 67−57 Mya in the stem lineage shared by the Podococceae, Oranieae, and Sclerospermeae (POS) clade and the core arecoid clade. The POS clade diverged into Africa before 43 Mya.

In the second part of the study (Baker & Couvreur, 2013b), the authors discussed the diversification history and origin of regional assemblages (see Fig. 6.5). The material and methods employed by the authors were the same as used in the first part of their paper (Baker & Couvreur, 2013a). The authors, however, employed a couple of new programs to assess the net diversification and extinction rates.

Their main inferences as relevant to our study are given here. (1) Thirty-five dispersal events took place between the regions. (2) The Indian Ocean and Pacific Ocean were the most frequent sinks (eight lineages each), four events (sinks) each were recorded into South America and Africa, and one

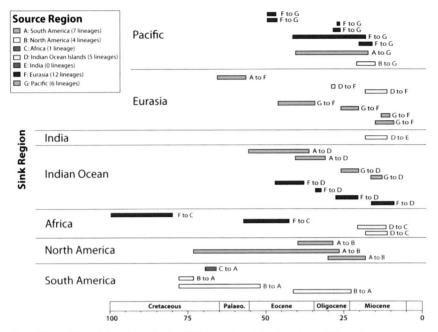

Figure 1 Temporal history of dispersal events in palms. All 35 dispersal events inferred from the most likely ancestral area reconstructions determined by LAGRANGE are illustrated, grouped into sink regions (see Fig. 1 of Baker & Couvreur, 2013 and raw LAGRANGE output at the Dryad data repository: doi:10.5061/dryad.vb25b35j). Each event spans the stem and crown node age of the branch along which it was inferred.

Figure 6.5 Spatial and temporal history of palm dispersals. *Reproduced with permission from Baker, W. J., & Couvreur, T. L. P. (2013b). Global biogeography and diversification of palm sheds light on the evolution of tropical lineage. II. Diversification history and origin of regional lineages.* Journal of Biogeography, 40, 286–298.

into India. (3) Though Laurasia (as Eurasia plus North America) was the most likely ancestral area for the crown node of palms, subsequent back-dispersals into these regions appeared to have been important with seven dispersals into Eurasia and three into North America. Eurasia was the most important source of dispersing lineages (12 nos.), followed by South America (seven lineages), the Pacific Ocean (six lineages), Indian Ocean (five lineages), North America (four lineages), Africa (one lineage), and India (none).

The authors had studied also the diversification rates of all the palm genera (see Fig. 6.6). Among the seven genera showing highest diversification rates, three belonged to the tribe Cocoseae—*Bactris, Butia,* and *Syagrus*. When high extinction rates were also factored, one out of the total six genera belonged to Cocoseae, viz., the genus *Bactris*. Regarding species

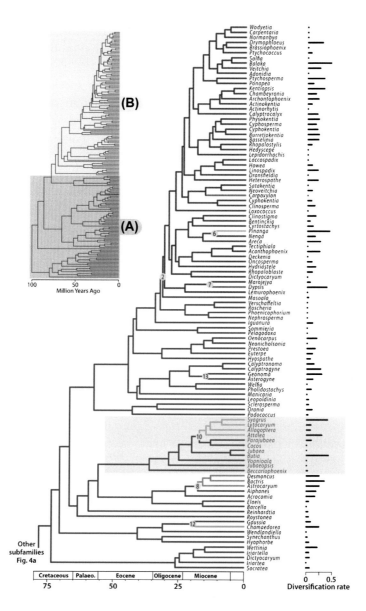

Figure 4 Diversification dynamics across palms. Chronogram of palms representing the 14 different diversification rates inferred across palms using TurboMEDUSA with the location of the 13 rate shifts increase. The figure is divided into two sections: panel (a) covering subfamilies Calamoideae, Ceroxyloideae, Coryphoideae and Nypoideae, and panel (b) covering subfamily Arecoideae only (see summary tree included in panel (b)). No rate decreases were detected. Numbered nodes indicate the location of rate shifts with a colour change indicating the lineages in which the shift has occurred. Bars next to generic names represent diversification rates under no extinction ($d = 0$). Estimates for the net diversification (r) and the relative extinction rate (epsilon) for each clade identified using turboMEDUSA are given in Table 1.

Figure 6.6 Diversification dynamics of subfamily Arecoideae palm. *Reproduced with permission from Baker, W. J., & Couvreur, T. L. P. (2013b). Global biogeography and diversification of palm sheds light on the evolution of tropical lineage. II. Diversification history and origin of regional lineages.* Journal of Biogeography, 40, 286–298.

diversity (under the constant birth/death diversification model), one of six genera belonged to Cocoseae. It was *Bactris* again.

Baker and Couvreur (2013b) attributed the high species diversity in South America to local diversification and not immigration. Most of the palm diversity present in South America happened in a single Cretaceous (144–66 Mya) dispersal event, with the remaining three dispersals accounting for only modest increases in species numbers.

Regarding the tribe Arecaceae, South America was the major source of dispersal. Nevertheless, the authors were unable to explain the phenomenon of long-distance dispersal of the tropical groups. They suggested a possible dispersal route of megathermal taxa via Antarctica that had happened mainly during the climatic optima of the Paleocene (66–57 Mya) and the Miocene (23–5 Mya). The lineages then began to diverge when the climatic barrier was reinstated. The dispersals from South America to the Indian Ocean and the Pacific appeared to have occurred during 55–40 Mya.

Baker and Couvreur had identified three components in the Neotropical palm flora (of Central America and the Caribbean). In the second component, three dispersals took place from South America to North America during the Cenozoic (1.6–0.01 Mya), and this included some genera of the tribe Cocoseae as well.

The authors indicated further that the present relatively small palm flora of India (c. 90 species) was more likely because of relatively recent colonization. They agreed with Dransfield et al. (2008) that the presence of Cenozoic palm fossils in west-central India indicated ancient dispersals. At the same time, Morley (2000) had suggested that the rapid latitudinal change in India's position from Early Cretaceous (100–145 Mya) until its collision with Eurasia about 35 Mya (Ali & Aitchison, 2008) might have caused considerable extinctions (Prebble & Dowe, 2008). In contrast, the dispersals into Eurasia, the Indian Ocean, and the Pacific (7–8 events each per region) occurred at short intervals during the Early Eocene to Late Miocene. Baker and Couvreur (2013b) had inferred as many as 18 dispersals during this period.

The authors have themselves cautioned about the several limitations of this study. The most noteworthy ones among them have been listed at the beginning. Nevertheless, this is to date the most comprehensive study of palm biogeography. However, this needs to be compared and collated with certain sectoral studies carried out at various levels of classification by other authors.

Baker and Couvreur (2013a, 2013b) found that subtribe Elaeidinae was sister to subtribe Bactridinae in all the analyses. Within the subtribe Attaleinae, *Lytocaryum* was positioned as sister to *Syagrus* in maximum parsimony (MP) and ML analyses with high support and embedded within *Syagrus* with B-analysis. Incidentally, Noblick and Meerow (2015) have proposed the merger of *Lytocaryum* with *Syagrus*. This analysis, too, had positioned *Cocos* as sister to *Attalea*. The stem node of *Cocos* and *Attalea* was optimized ancestrally in the Atlantic coastal forest region. They have concluded that this study has provided the most accurate evolutionary history of the tribe Attaleinae to date.

Meerow et al. (2014) performed three BEAST runs with version 1.8.0. as a continuation of their earlier studies (Meerow et al., 2009). All of them gave the same tree topologies. Molecular clock was assessed using the likelihood ratio test of the program Hyphy 2.2.0. For calibration, the authors selected the fossil described from the Paleocene of southern Argentina by Futey, Gandolfo, Zamaloa, Cuneo, and Cladera (2012). Meerow et al. (2014) performed biogeographic analysis that presents a divergence history of Cocoseae dominated by dispersal events, which outnumbered vicariances three-fold through the entire chronogram (Fig. 6.10). Great diversification of palms occurred in the Paleocene, especially in northern South America.

The mean crown age of the tribe Cocoseae was 63.8 Mya (62.0–68.2 Mya). This was also the stem age of the subtribe Attaleinae clade. Attaleinae was the oldest of the three subtribes of Cocoseae, with a crown age of 37.8 Mya. The major lineages within each subtribe originated during the Oligocene (33.9–23.0 Mya). Within the entire tribe, speciation within each genus took place earlier than the Miocene (<23 Mya). Incidentally, Morley (2000) had observed that terminal Eocene cooling had little effect on the flora (including palms) of S. America.

The authors carried out further the biogeographic analysis by optimizing ancestral areas on the Bayesian supermatrix tree with S-DIVA program. However, they indicated that their inferences should be viewed with some caution, as their material had constituted only less than half the total number of species of some genera. Area optimization showed that the ancestral American Attaleinae was restricted to the Andes, and eastern and central Brazil during the Eocene–Oligocene boundary. This period was marked by severe cooling and Antarctic ice growth. Meerow et al. (2014) found that "the crown node of the American Attaleinae (P = .33) was optimized ambiguously at the Atlantic Coastal Forest region with both the Andean and Central Brazilian region with subsequent vicariance between the Atlantic Coastal Forest and Central Brazil–Andes regions".

The authors found that divergence of *Cocos* and *Attalea* and the earliest diversification of *Syagrus* coincided with the early Miocene divergence between *Parajubaea* and *Allagoptera* and that the earliest subsequent diversification within the tribe Cocoseae was attributable to the Andean uplift that occurred during the Late Miocene–Pliocene period. The ancestral area of the stem node (P = 1) for the *Cocos, Attalea,* and *Syagrus* (with *Lytocaryum* embedded) subclades was optimized in the Atlantic Coastal Forest with stasis (the persistence of a species over a span of geological time without significant change; King et al., 2006) until dispersal occurred to the Amazon at the crown node of the "rain forest clade" of *Syagrus* (P = .1). The stem node of *Cocos* and *Attalea* was optimized ancestrally in the Atlantic Coastal Forest Region and the crown node of *Attalea* too, had an ancestral optimization in the Atlantic Coastal Forest Region. From the crown node of *Cocos*, 14 dispersals were predicted to account for the "broad terminal distribution" of the genus. The BEAST supermatrix analysis supported monophyly in this analysis except in *Syagrus* (Meerow et al., 2014) (see Fig. 6.7).

The present study has shown some notable differences in the resolution of the American Attaleinae from the earlier accounts. We shall cover only those aspects that are relevant to the coconut. Although *Cocos* was still placed as sister to a *Lytocaryum/Syagrus* subclade in the MP and ML supermatrix analyses, its support was less than that in Meerow et al. (2009). "Two of our species–tree estimation analyses position *Cocos* as sister to *Attalea*, as does the B-analysis of the supermatrix, although never with strong support. The present ML species tree agrees with the MP and ML supermatrix topology" (Meerow et al., 2014).

Meerow et al. (2014) further tested the sister relationship of *Cocos* and *Attalea* as a constraint on the supermatrix with MP. They found that "only four additional base substitutions" were adequate to support the resolution. They stated further that "like *Attalea*, *Cocos* has a fibrous mesocarp, which is unlike the fleshy mesocarp of *Syagrus* and *Lytocaryum*…. Single gene analysis of *WRKY* 6 and *WRKY* 21 resolved *Cocos* as sister to *Attalea*. Only one gene tree, *WRKY* 19, resolves *Cocos* as sister to *Syagrus* and *Lytocaryum*." The present data also did not support the position of *Cocos* as sister to *Parajubaea,* as observed by Gunn (2004) and Baker et al. (2009). "We are inclined to accept the sister relationship of *Cocos* and *Attalea* as the more likely scenario, given the consensus of two-species–tree diversification within *Syagrus* coinciding with the formation of the Pebas Sea as a physical barrier between eastern South America and the Andes. This lasted till late Miocene."

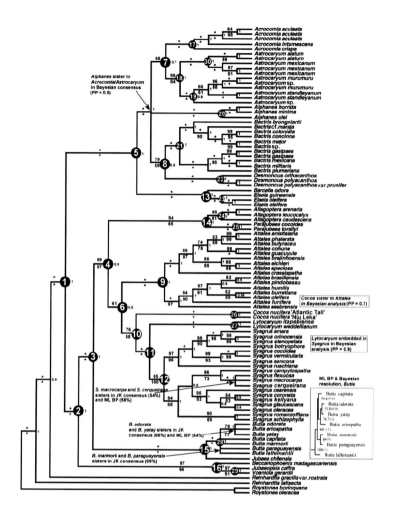

Fig. 1. Strict consensus tree of 12 equally parsimonious trees found by heuristic parsimony analysis of six combined WRKY loci sequences aligned across Areceaceae tribe Cocoseae, with *Roystonea* and *Reinhardtia* as outgroups (*Roystonea* is the functional outgroup). Numbers above branches are parsimony jackknife support percentages. Numbers below branches are ML bootstrap (BP) support (non-partitioned analysis). Numbers at nodes are posterior probabilities from the non-partitioned BEAST analysis. The numbered nodes are referenced in Table 2 (partitioned decay indices). The inset box shows the ML resolution of *Butia* with BP percentages. * = 100% support.

Figure 6.7 Strict consensus of 12 equally parsimonious trees across tribe Cocoseae. *Adapted with permission from Meerow, A. W., Noblick, L., Salas-Leiva, D. E., Sanchez, V., Francisco-Ortega, J., Jestrow, B., et al. (2014). Phylogeny and historic biogeography of cocosoid palms (Aracaceae, Arecoideae, Cocoseae) inferred from sequences of six WRKY gene family loci. Cladistics, 1–26. http://dx.doi.org/doi:10.1111/cls.12100.*

Meerow et al. (2014) had further studied the differentiation of the entire tribe Cocoseae (see Fig. 6.8) They had used also more distantly related taxa as outgroups [as against only subtribe Attaleinae in their earlier study (Meerow et al., 2009)]. They employed several new programs in their attempts to get

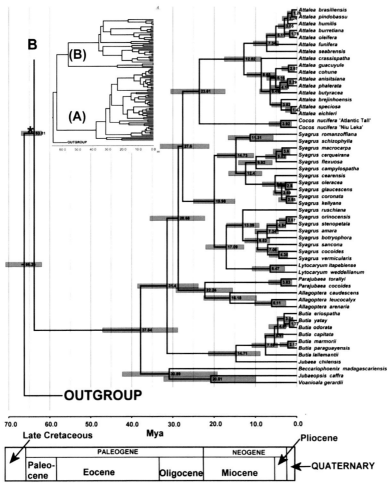

Fig. 2. Maximum clade consensus chronogram of 15 000 trees sampled from non-partitioned Bayesian analysis of six combined WRKY loci sequences aligned across Arecaceae tribe Cocoseae. Numbers at nodes are estimated ages (Ma) of clades; the grey bars are 95% highest prior density indices. * = calibration point. (a) Outgroups and subtribe Attaleinae; (b) subtribes Bactridinae and Elaeidinae. Posterior probability scores are reported in Fig. 1 for all nodes congruent with the MP and ML trees.

Figure 6.8 Maximum clade consensus chronogram obtained from nonpartitioned Bayesian analysis aligned across tribe Cocoseae. *Adapted with permission from Meerow, A. W., Noblick, L., Salas-Leiva, D. E., Sanchez, V., Francisco-Ortega, J., Jestrow, B., et al. (2014). Phylogeny and historic biogeography of cocosoid palms (Aracaceae, Arecoideae, Cocoseae) inferred from sequences of six WRKY gene family loci. Cladistics, 1–26. http://dx.doi.org/doi:10.1111/cls.12100.*

better resolution in their results. Attaleinae was the oldest of the three sub-tribes with a crown age of 37.8 Mya. Overall, the major lineages within each subtribe originated during the Oligocene (33.9–23.9 Mya). The authors felt that the divergence history of the entire tribe Cocoseae was dominated by dispersals, which had outnumbered vicariances three-fold (see Fig. 6.9).

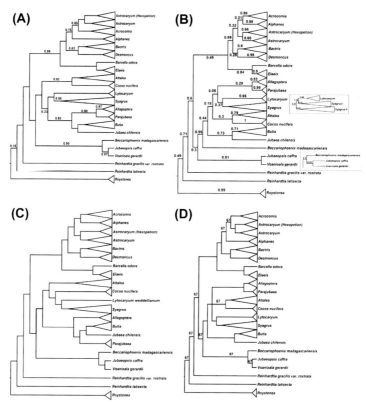

Fig. 3. Results of three gene-tree/species-tree estimation approaches on the phylogeny of the palm tribe Cocoseae using six WRKY transcription factor loci, and 50% majority rule consensus of supermatrix MP tree and two species trees. (a) Maximum clade credibility consensus from partitioned *BEAST analysis using BEAST Version 1.75. Numbers above or below branches are posterior probability (PP) scores. Only PP < 1.0 are shown. (b) Population tree (resolved by quartet solutions) from BUCKy analysis (Larget et al., 2010), with concordance factor scores above or below branches. Insets are resolution of two nodes from the primary CF tree (resolved by highest CF scores) that differed from the population tree. (c) Highest scoring ML species tree inferred from six WRKY loci consensus gene trees using STELLS. (d) Fifty per cent majority rule consensus tree of one of 12 equally parsimonious trees found by MP, the *BEAST species tree, and the BUCKy population tree. Only consensus indices < 100% are shown, above or below branches.

Figure 6.9 Three gene–tree/species–tree estimation approaches on the phylogeny of tribe Cocoseae. *Reproduced with permission from Meerow, A. W., Noblick, L., Salas-Leiva, D. E., Sanchez, V., Francisco-Ortega, J., Jestrow, B., et al. (2014). Phylogeny and historic biogeography of cocosoid palms (Aracaceae, Arecoideae, Cocoseae) inferred from sequences of six WRKY gene family loci.* Cladistics, *1–26. http://dx.doi.org/doi:10.1111/cls.12100.*

Cocoseae is rich in animal–dispersed fruits (Meerow et al., 2014). They stated further that an extinction event occurred only once. The origins of the tribe were dated to Late Tertiary, by which time, direct connection between Africa and South America had been severed already, and the asteroid impact at the Cretaceous–Tertiary boundary, at c. 65.5 Mya, had resulted in extinction of nearly a third of terrestrial vegetation and a great decline of species abundance (Nicols & Johnson, 2008). The authors opined that the big gap between the stem (63.8 Mya) and crown (37.8 Mya) nodes of Attaleinae might be a consequence of this event.

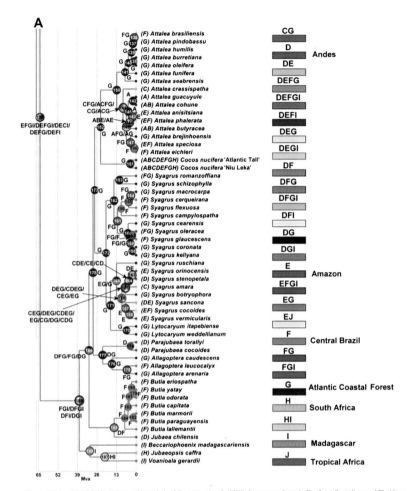

A

Fig. 4. Results of the S-DIVA biogeographic analysis of Cocoseae using the MCC chronogram shown in Fig. 2. (a) Bactridinae and Elaeidinae. (b) Attaleinae. The ten regions assigned to the 96 terminal taxa in our matrix were the seven American palm regions enumerated by Henderson et al. (1995) with the inclusion of three Old World locations. Refer to Table S3, Supporting Information for the event probabilities and putative routes. Nodes with red circles around them have inferred dispersals, blue circles = inferred vicariance, green = extinction event. The nodes are labelled with ancestral area codes as per legend.

Figure 6.10 Results of S-DIVA: a biogeographic analysis of tribe Cocoseae, subtribe Attaleinae (unarmed species only). *Reproduced with permission from Meerow, A. W., Noblick, L., Salas-Leiva, D. E., Sanchez, V., Francisco-Ortega, J., Jestrow, B., et al. (2014). Phylogeny and historic biogeography of cocosoid palms (Aracaceae, Arecoideae, Cocoseae) inferred from sequences of six WRKY gene family loci. Cladistics, 1–26. http://dx.doi.org/doi:10.1111/cls.12100.*

Meerow et al. felt that a great diversification of palms appeared to have occurred during the Paleocene (see Fig. 6.10), particularly in northern South America. This has also been observed by several earlier workers. The subtribe Attaleinae was absent in Central America until the late Neogene entry of *Attalea*. The African presence of Attaleinae is related to three "reliefs." They proposed that the two possible routes for this vicariance might have been by either (1) migration via the Antarctica during the Eocene thermal maximum, or (2) dispersal via South Africa where there were rain forests during Eocene (Morley, 2000). The evidence supporting this could be the recovery of Eocene fossils of *Cocos*-like plants from New Zealand and India. They felt further that the early Miocene flooding of western Amazonia caused by the uplift of the Eastern Cordillera of the central Andes, and subsequent Caribbean marine incursion (Pebas Sea) appeared to coincide with the divergence of *Attalea* and *Cocos* and also their earlier diversification from *Syagrus*. Further, much of the subsequent diversification within Cocoseae was attributable to the Andean uplift that occurred during Miocene to Pliocene (Kissling et al., 2012).

Meerow et al. (2014) then concluded that "as regards the shifting landscape of the relationships of *Cocos*, the sum total of available evidence, including individual gene tree resolution, compels us to accept the sister relationship between *Cocos* and *Attalea*."

We had covered in the last chapter the details of the molecular biology studies carried out by various authors and the discussions that followed each study. Hence, we have confined here only their findings related to biogeographical aspects.

The genus Attalea is one of the three genera that have been found to have the closest relationships with the genus Cocos. Freitas et al. (2016) conducted phylogenetic studies in the genus Attalea to get insights into the historical biogeography of the genus. Presently, the genus includes about 79 species. They are distributed from Mexico to Bolivia and Paraguay. They occur from lower slopes in the Andes up to 1800 m msl, characterized by high ecological variation. The formerly recognized groups within Attalea were not found to be monophyletic. The species is acaulescent, or stemless, to erect but is always erect and solitary.

The authors used 79 accessions of 49 species in their studies. Their study suggested that the common ancestor of Attalea was present in the Atlantic coastal forests c. 30 Mya. It confirmed monophyly of the genus. The three main clues in the dendrogram appeared to correspond to the previously recognized genera—Scheelia, Orbignaya, and Maxmilliana. They showed as ecogeographic biomes. The authors then concluded that the entire diversification within Attalea occurred in relatively recent times.

CHAPTER 7

Origins: Past Observations

1. INTRODUCTION

The credit for transforming the inquiries on crop-plant origins into a scientific discipline should go to the French botanist, Alphonse de Condolle (1806–1893), who is best known as the author of the publication, Origin of Cultivated Plants (1882, 1886). De Condolle was the first person to use evidences from fields as diverse as history, archeology, folklore, prehistory, taxonomy, botany, and travelogs, to draw inferences on crop-plant origins. In his book, De Condolle discussed the origins of more than 300 crop plants. This also included the coconut. Since then, several authors have reviewed and discussed the origins of the coconut. However, mainly, almost all of them are speculative and repetitive in nature.

Nevertheless, only a few authors have discussed this problem from different perspectives: Cook (1901), Mayuranathan (1938), Sauer (1971), Whitehead (1974), and Harries (1978), for instance. Hardly any experimental studies have been done on this subject so far. The main reason appears to be the perennial habit of the coconut palm, its monospecific nature (according to the current classification of the genus *Cocos* L.), and most importantly, lack of funding support for taking up such studies.

Any inquiry into the origins of a cultivated crop plant should involve discussions on four aspects components—time of origin, place of origin, the ancestral species/taxon, and mode of speciation.

2. REVIEW OF LITERATURE

2.1 De Condolle's Views

De Condolle (1886) began his discussion on origin of the coconut by mentioning about the sightings of coconuts (1) in the 16th century by the famous palm scientist, von Martius (1794–1868) in Central and South America—in the provinces of Bahia and Pernambuco (eastern Brazil); (2) in the then uninhabited island Cocos Island (now in Costa Rica, 300 km west of Panama) by the navigators, Dampier and Vancouver; and (3) by the botanists, Seemann and others, on the isthmus of Panama, of both wild and cultivated

The Coconut
ISBN 978-0-12-809778-6
http://dx.doi.org/10.1016/B978-0-12-809778-6.00007-3

117

coconuts. He then presented arguments both in favor and against the origin of coconuts in America and Asia. And, finally, he advanced historical information about the antiquity of the coconut in South, Southeast, and East Asia. He concluded, "I formerly thought that the arguments in favour of southern America were the strongest. Now, with more information and greater experience in similar questions, I incline to the idea of an origin in the Indian archipelago. The extension towards China, Ceylon, and India dates from not more than three or four thousand years ago, but the transport by sea to the coasts of America and Africa took place perhaps in a more remote epoch, although posterior to those epochs, when the geographical and physical conditions were different to those of our day" (De Condolle, 1886).

De Condolle's observations indicate two things—that coconuts originated in the Indian Ocean archipelago; and that the stands of the coconuts found in some regions along the South American coast and in a couple of islands off the American coast were the result of floatation and establishment in the remote past. He did not state anything about the other aspects of origin of the coconut.

2.2 American Origin of the Coconut

O.F. Cook (1901, 1910), then a Botanist at the US National Herbarium, forcefully rebutted the widely prevalent belief that the coconut originated in the western Pacific, Indian Ocean, or Southeast Asia. In his lengthy and elaborate narrative, he first contradicted the 10 points advanced by De Condolle (1886) favoring the Indian archipelago as the place of origin of the coconut. He was emphatic that there were no instances of establishment of coconut populations anywhere in the world, after dissemination of the nuts by water currents, and that every stand of the palm that has been found thus far on land and/or on islands has been due only to human efforts. Coconuts were already present in the Americas at several locations by the time Columbus landed in Hispaniola in 1492. Columbus himself had observed coconut stands in Cuba during his first voyage (not in the fourth voyage, as widely reported), Cook stated. The author then mentioned the sightings of coconut stands at several locations on the west coast of Central and tropical South America, the most notable being on the Isle of Coco (about 300 km west of Panama). The coconut was present also on the Atlantic side of the American tropics, as for instance, in Cuba, Puerto Rico, Brazil, and Colombia, Cook claimed.

It is now generally accepted that most of these sightings were unlikely to have been of the coconut, but of other palm species that possess the gross morphological features of the coconut palm. This is because most of these early travelers would not have been sufficiently familiar with the morphology of the tropical coconut palm, that they would have been able to distinguish it as real coconut palm from a distance.

In further support of his contention of the American origin of the coconut, Cook pointed out that all the palms that were related to the coconut—comprising about 20 genera and 200 species (it was so in his time; Author)—were natives of the Americas, with the possible exception of a single species, the west African oil palm (*Elaeis guineensis*). Further, all the species of the genus *Cocos* that were the most related to the coconut were natives of the interior valleys and plateaus of the Andes, where the coconut also thrived, remote from the sea, Cook claimed. In another place in the text, Cook (1910) stated unequivocally that the coconut "must have been a native of South America and carried westward across the Pacific in prehistoric times … and its original home must be sought in some sheltered valley of the equational[sic] Andes" (eastern valley of Peru), according to the author.

Cook then went on to state that out of 18 *Cocos* species, all except *Cocos nucifera* were natives of Brazil (in Griesebach, 1864; Flora of the British West Indies, cf. Cook, 1901), and this indicated that the "original habitat" of the coconut was the west coast of Panama. Cook argued that all the close relatives of coconut were natives of relatively dry interior regions of Brazil, and coconut too was adapted to grow in such habitats. He was clear that the smooth, thick epicarp of the fruit, the dry fibrous mesocarp, the thick hard endocarp, and the water inside, were all adaptations for survival and establishment under low-moisture conditions.

This point of view is contrary to the widely held view that the above characteristics of the fruit are in reality adaptations for dissemination of the species by water.

The second author, who gave strong support for the origin of coconuts in South America was the Norwegian seafarer and explorer, Thor Heyerdahal (1914–2002), of Kon-Tiki expedition fame (Heyerdahal, 1950) (Figs. 7.1 and 7.2).

To prove his contention that it was the American Indians who were the original colonizers of the Pacific Ocean islands, Heyerdahal, along with five other navigators, set sail across the Pacific to the west on a raft made of nine logs of balsa wood [*Ochroma pyramidale* (Lam.) Urb., Malvaceae] using only the prevailing winds and ocean currents. They set sail from the Peruvian port town of Callao (12°S, 77°W) on April 28, 1947. After more than

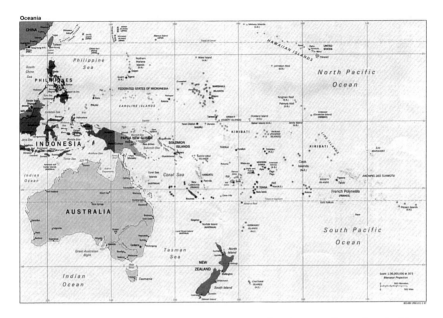

Figure 7.1 Pacific Ocean, showing places mentioned in the text. *Reproduced with permission of the University of Texas Library.*

3 months voyage, on August 7, as the raft passed south of the Marquesas Islands (9°30′S, 140°0′W) and was approaching Tahiti, it ran aground on Raroria reef (6°42′S, 99°42′W) in the Tuamotu Archipelago (all in French Polynesia). Heyerdahal and his associates had sailed in the raft for 45 days and covered more than 6000 km from Peru (Heyerdahal, 1950). The author claimed that possibly they could have continued to sail further "by making the steersman stand and pull a center board up and down in a chink, instead of hauling sidewise on the ropes of the steering oar." But, they did not do this because they had proved their point that it was possible for humans to sail across from the American coast by using just the power of the ocean currents and winds.

2.3 The Opposing Views on American Origin

Incidentally, there have been always lingering suggestions of Pre-Columbian contacts between the Old and New Worlds. Several domesticated plants and animals are known to have been exchanged between the two worlds. In particular, there is general agreement that the sweet potato, bottle gourd, dogs, rats, and chickens have been exchanged between South America and the Pacific islands. Some other plants that are in this category include the

Indian Ocean Area

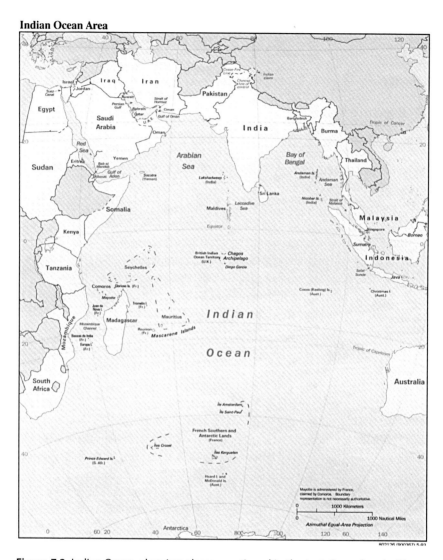

Figure 7.2 Indian Ocean, showing places mentioned in the text. *Reproduced with permission of the University of Texas Library.*

banana, coconut, squash, corn, and *Phaseolus* (Nayar, 2010; Riley, Kelley, Pennington, & Rands, 1971).

Storey et al. (2007) obtained a radiocarbon date of 1304–1424 CE of a DNA sequence taken from an ancient chicken bone found in an archeological site in the Aranco Peninsula, Chile. This indicated a Pre-Columbian introduction of chicken to the Americas from Polynesia, where chickens had been introduced earlier by the Austronesians from Asia. Readers are invited to refer to Riley et al. (1971) and Jones, Storey, Matisoo-Smith, and Ramirez-Aliaga (2011) for detailed updates on Pacific Ocean exchanges.

2.4 Beccari's Views

Then, Beccari (1917) came out with a strong rebuttal of Cook's (1901, 1910) propositions. He observed that the three principal theses with which he entirely disagreed, and on which Cook had built up his argument, were the following:

(1) The coconut assumed its present characters on the South American continent. The Polynesian navigators discovered coconuts in the Americas and they disseminated it in Polynesia and then on to Asia. (2) The coconut is unable to spread and establish anywhere without human protection. (3) Ocean currents could not have been an efficient means of dispersal of the coconut.

Beccari countered the first argument by citing the presence of *Elaeis* in Madagascar and *Jubaeoposis* in South Africa. He stated also that *Jubaeopsis caffra* had "many more affinities with *Cocos nucifera* than any other palm." To counter the second point, Beccari stated that the giant robber crab, *Birgus latro*, has a very close (symbiotic) relationship with the coconut, and this crab species was not known to be present in the Americas. Regarding the third point, Beccari contested the contention of Cook that the Portuguese or Spanish could not have been expected to introduce the coconut to the Americas because the nuts would not have been able to retain their viability during their voyages from Europe. Beccari cited also the wide occurrence of coconuts on several Pacific Ocean islands including the atolls. He mentioned the prevalence of a longstanding practice in Malaysia of tying coconut bunches in the interiors of living places, and also of the nuts germinating while hanging in the rigging of a ship. In a similar vein, Beccari (1917) went on to counter the various other agreements of Cook (1901, 1910) favoring the South American origin of coconuts.

The rationale of Beccari (1917) to state that *Jubaeopsis caffra* was the species most closely related to the coconut is unclear. It was Beccari who had erected both the genus *Jubeopsis* and the species *J. caffra*. It is a clustering palm with short erect stems that frequently branch dichotomously. It grows gregariously on the coastal beaches of two rivers in Pondaland, South Africa (Dransfield et al., 2008). It is endemic to this region. It is monoecious. Its fruit is c. 3-cm long, and is one-seeded. Its chromosome number is $2n = 160–200$ (coconut, $2n = 32$).

The inferences of Beccari (1917) received support from another countryman–botanist of the time, Chiovenda (1921–1923), who also disagreed with a westward dispersal of the coconut from South America.

Incidentally, recent molecular biology studies support the radiation of the progenitors of the coconut from South America to the Pacific. Further, Dransfield, Flenley, King, Harkness, and Rapu (1984) had recovered from a cave in Easter Island (27°00′S, 109°00′W) about 30 partially gnawed endocarps of a cocosoid palm, now extinct. It was designated, *Paschalococcus disperta*. The fruits measured 3.0×2.5 cm, and it was carbon-dated to 820 ± 40 year BP. They observed that *P. disperta* fruits looked similar to those of *Jubaea*. *Jubaea* is a monotypic species, belonging to the subtribe Attaleinae. *J. chilensis* (Chilean wine palm) is a massive palm. It is restricted in its distribution to central Chile. It is now a threatened species, as it used to be felled wantonly at one time for tapping the stem for wine and sugar (Dransfield et al., 2008).

Beccari published two significant papers (Beccari, 1916, 1917). In the first one, Beccari shifted all the species of the genus *Cocos* to other genera such as *Syagrus and Butia*, except the coconut. This left *Cocos* as a monotypic genus with only *Cocos nucifera* included in it. Its relevance may be noted from the fact that all the palm taxonomists subsequently have endorsed this change.

2.5 The Indian Ocean–Western Pacific Ocean Region as Center of Origin

In the second publication, Beccari (1917) had dealt with the origin and dispersal of the coconut. We have previously covered this work. Beccari then cited several examples of islands and atolls that were uninhabited by humans where natural stands of the coconuts occurred, including the isolated Palmyra Islands (5°52′N, 162°5′W). He then stated that *Jubeopsis caffra* appeared to have much more affinity with *Cocos nucifera* than any other palm species that authors had hitherto referred to the genus *Cocos*. As for the place of origin, Beccari indicated that it "may have originated not in some lands which have now disappeared from the Pacific, but in islands lying in the eastern Indian Ocean, or in some other lands or islands existing in former times between Africa and India … Ceylon and Keeling (now, Cocos Islands) Islands must lie almost in the region where *Cocos nucifera* assumed its present characters. The coconut, he concluded, "is a halophilous plant with a predilection for the seashore."

Mayuranathan (1938) analyzed the then-available information on the original home of the coconut covering the areas of folklore, religious beliefs, linguistics, and geology. He observed that the coconut formed one of the most used plant items in the Hindu religious rites of peninsular India since about the time of *puranas*.

Unlike the other ancient civilizations of China, Greece, Rome, and the Middle East, the written script came late in the Indian civilization. The texts of all the ancient scriptures like the Vedas, Puranas, Mahabharata, and Ramayana were being carried down the ages through oral traditions till the development of the written script.

According to the noted early-period Indian historian, Romila Thapar, early archeology in the Indian subcontinent (c. 1200–600 BCE) had revealed the existence of many diverse cultures, mostly chalcolithic, either interwoven or in juxtaposition. Because evidence of the material culture of the late second and early first millenium BCE is relatively clear, there is a large range of texts with varying narratives and different dates. Those that constituted the vedic corpus belonged to this period. They began as oral tradition memorized with much precision, which was eventually written down many centuries later. Mahabharata, Ramayana, and Puranas also began as oral traditions. They were converted to the present textual form only in the early first millennium CE. Puranas are later than the Vedas (Thapar, 2002). Puranic accounts narrating the beginnings of Indian history are largely variations of a well-known theme, as narrated in Vishnu and Matsya Puranas (Thapar, 2002).

The coconut is mentioned in all the numerous puranas (Thapar, 2002). Singh (2009) has indicated 50 BCE–550 CE as the period when the Puranas were written down. The oral traditions were much older, of course.

Mayuranathan observed that coconut did not constitute one of the 16 native plants then being used in after-death rites. His source of this information is not given. This is not, however, the case in the Konkan–Malabar coastal region of peninsular India. He felt that the early historic and linguistic evidences indicated that many of the terms and names used for the coconut in the vast stretch from the Marquesas Islands (northern French) Polynesia (9°S, 140°W) to Madagascar (east of southern Africa) appeared to have been derived from the early Sanskrit language.

Mayuranathan (1938) felt that the original home of the coconut should be a region where the nuts would freely germinate without watering, where its cultivation would demand the least care, where the plant could provide for itself, and where it would grow very prolifically, as is the case in Malaya, the Eastern Archipelago, or the islands of the west and central Pacific Ocean in the equatorial belt, where the coconuts grow "as densily[sic] as bristles in a brush." The original form of the coconut must have been one with viviparous adaptations, and the inland form of today should be a retrograde (sic, primitive?) from the original, he added.

Mayuranathan stated further that, though the genus *Cocos* was American (according to the then-prevalent classification), much of the tropical western South American region was generally dry and that there was no mention about the use of coconut by the native people of the South American continent. The wide presence of naturally occurring coconuts in several Indian Ocean islands, and other factors mentioned earlier indicated to him that the coconut was Southeast Asian in origin, and not South American. The author pointed out the division of the far-eastern region of Southeast Asia into three zones—the regions that are now known as Sunda, Sahul, and Wallacea (Fig. 7.3). Mayuranathan then concluded that biological evidence

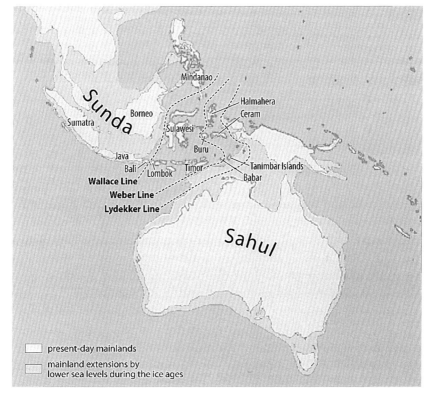

Figure 7.3 Wallace line, Sunda, and Sahul. *Adapted from Green, R. C. (1991). Near and remote Oceania. In A. Pawley (Ed.), Man and a half (pp. 491–502). Auckland: The Polynesian Society.*

proved that there were older lands connected with New Guinea that are now entirely submerged "consisting of the sea shelf on the northwest of New Guinea in the region now occupied by Geelvink Bay, or the area just to the north of it…" (this region is now known as Teluk Cenderawasch in Papua, presently; 2–3°S, 134–138°E). The author did not indicate how he arrived at this inference.

Sauer (1967, 1971) studied extensively the plant life of the Seychelles Archipelago and the West Indies. The Seychelles group of islands is located in the southwest Indian Ocean, north of Madagascar (5°S, 56°E) (Fig. 7.2). They comprise a compact group of mountainous islands (115 nos.; 455 km² area) rising steeply from the sea and another group of low-lying coral islands and atolls to their southwest, that are flat, waterless, and mostly uninhabited. It is assumed that "geologically, the whole Indian Ocean islands might have emerged as a single island during the last Pleistocene glaciations with hundreds of times the present size" (Wikipedia, 2015). In modern times, the first ship to touch the Archipelago was in 1609 by the British East Indiaman, the Ascension (Wikipedia, 2015).

Sauer observed that the extensive occurrence of coconuts in the Seychelles Archipelago had been mentioned in the accounts of the medieval age travelers like Ibn Batuta [1304–1368/1369 CE (cf. Gibb, 1929)], Gift in 1929, Sulaiman (9th century CE), etc. In addition, during the Middle Ages, when coastal trading by the Arab seafarers was common along the northern Indian Ocean regions (western Indian, Arab and Persian Gulf, east Africa, and the Red Sea), several of the ships used to be made here entirely of the coconut, complete with rigging and sails. The author felt that this signified the mastery of ship building and sailing of the local inhabitants from at least the early centuries of the Common Era, and the importance of coconuts in their lives. Sauer had no doubt about the coconuts having been a component of the native flora, and also as strand vegetation by the floating of coconuts across the ocean in some islands. He was certain that the vegetation was "certainly virgin and natural," and the coconuts were one of the components.

Sauer gave instances of the natural occurrence of coconuts in several other islands and atolls in the Pacific Ocean to strengthen his contention—Palmerston Atoll, Palmyra Atoll, Cocos Islands, etc. Then the author concluded "Thus, *Cocos nucifera* is best regarded as a semidomesticated species with a complex of local populations with all degrees of dependence on man from nil to complete. … At our present level of undertstanding[sic], the trans-Pacific distribution of the species is not reliable evidence of human dispersal."

2.6 Child, Purseglove, Whitehead, and Mahabale

Child (1974) reviewed the earlier literature and concluded that the trend of evidence favored the origin of the coconut in the Old World. "It must be admitted that we cannot, and are unlikely to be able to, assign to *Cocos nucifera* a precise habitat in the Pacific Ocean between 80°[sic]E and 180°[sic]E latitude. He opined that natural distribution by sea, though limited, can definitely occur." He cited as example the appearance of the coconut on the volcanic island of Anak Krakatoa IV (Hill & van Leeuwen, 1933) and also the reappearance of coconut on the Bikini Atoll in the Marshall Islands, 9 years after nuclear testing was done there in 1946, when all the vegetation had been killed (Williams, 1967). Child concluded that the "wide area under coconuts today is certainly largely due to the influence of man. Inland, every tree owes its existence to man, on the coasts, most of them do so."

Purseglove (1968, 1985) reviewed in detail the past literature on the origin and dispersal of the coconut. About dispersal, he opined that "it can be dispersed by ocean currents and can establish itself, although rarely, in coastal areas without the aid of man."

In support, he quoted the opinion of Ridley (1930), and highlighted the observations made in Krakatau, Anak Krakatau IV, and Bikini Islands. He mentioned also about the finding of Charles Darwin in 1836 that Cocos (Keeling) Islands were thickly covered with coconuts. Incidentally, Cocos Islands (12°10′S, 96°55′E, 14.2 km² area) is now administered by Australia and lies c. 2200 km northwest of it. It was an uninhabited island at the time of Darwin's visit. Dowe and Smith (2002) have attested to the presence of naturally occurring coconut in these islands in recent times.

The Krakatau volcanic island was formed in 1883 after a volcanic eruption. It is located 25 km south of Java (Indonesia). The Dutch and British scientists had visited these islands in 1886, 1897, and 1906 (Hill & van Leeuwen, 1933; Van Leeuwen, 1933). They found that the entire vegetation had been destroyed by the earthquake in this and the two nearby islands of Lang and Verlaten. All of them were covered with 30–50 m of pumice. In 1897, i.e., 13 years after the earthquake, coconut palms were already established in the islands. In 1906, the authors saw bearing coconut palms in the other two islands also, Krakatau and Verlaten. Anat Krakatau IV was another island that had emerged from the sea in 1930. Van Leeuwen (1933) found 41 germinating coconuts in this island in 1932.

Purseglove (1985) cited the opinion of Sauer (1967) that at least part of the coconuts in the Seychelles Islands could have arisen by natural dissemination. Based on all the previous observations, he concluded that "coconuts can be carried by ocean currents and can establish themselves on open coasts without the aid of man, even if such establishment may be rare." He then supported the observation of Child (1974) that "the exceptionally wide area under coconuts today is certainly largely due to the influence of man."

Regarding the place of origin of the coconut, Purseglove (1985) stated that the "available evidence pointed to the domestication of the coconut in the Pacific area, a view that was held also by De Condolle (1886), Beccari (1917), Mayuranathan (1938), Vavilov (1950), Merrill (1954), Fosberg (1962), Zeven and Zhukovsky (1975), Child (1974), and Corner (1966). It is not necessary to invoke the aid of man in the transfer of plants between the New and Old Worlds, and vice versa before 1492." He then listed 12 reasons in support of this contention.

Most of these points have been already covered in the earlier reviews. A different perspective, however, merits mention here. Lepesme (1947), in his monograph on the insect pests of palms, observed that 74 species were specific to the coconut, and this represented as much as 47% of the total Melanesian fauna. Because everywhere else this percentage value was much lower, he felt that coconuts might have evolved in Melanesia.

Some of the observations made by Purseglove (1985) and ascribed to other authors, appear to be different from the observations actually made by those authors. For instance, Child (1974) had stated merely that coconut might have originated somewhere between 80°E and 180°E. He had stated further that it was difficult to assign a more precise location.

Whitehead (1974) dealt with the origin of the coconut in Simmonds (1974), Evolution of Crop Plants. He observed that the Atlantic/Caribbean coasts have talls similar to Jamaica Talls (except in Nicaragua, Venezuela, and Surinam), and, with few exceptions, coconuts on Pacific Islands resembled the Panama Tall more in major characteristics than the Jamaica Tall. He did not specify what they were. He further added (quoting Fremond, Ziller, & de Nuce de Lamothe, 1966; Whitehead, 1965, 1966) that the coconuts on the Caribbean/Atlantic coasts "broadly resemble those of west Africa" and these in turn "are essentially similar to those of east Africa, Seychelles, Mozambique, and Sri Lanka." From this, he inferred that "Atlantic–American palms arrived relatively recently (meaning post-1492) from Southeast Asia via Africa, or perhaps via the Seychelles." He did not give any supporting evidence for making this statement.

This, however, appears to be an oversimplistic generalization. Since the period for which information and data are available, it is well documented that most of the coconuts have been occurring in South and Southeast Asia from prehistoric times. However, increased movements of people including transfers of coconuts have been taking place from the late 15th century.

Firstly, the movement of people and plants from Southeast Asia to Madagascar has been taking place at least from the early centuries of the Common Era, taking advantage of the southwest and northeast monsoon currents (Murdoch, 1959; Sheriff, 1981; etc.) (Figs. 7.4 and 7.5). In the literature, there does not appear to be any mention of these boats touching the Seychelles, though some occasional landfalls might have taken place there also. Several of the boats used to touch South Asia en route.

Secondly, Randomly Amplified Polymorphic DNA (RAPD) analyses of coconut palm populations using 17 representative and distinct South Pacific coconut populations showed that about 60% of the observed diversity in the different populations occurred within the populations themselves, and that this varied among populations (Ashburner, Thompson, Halloran, & Foale, 1997). The average diversity among populations was low; it ranged from 0.05 to 0.28. Although the populations of the region displayed continuous variation, the 17 populations could be divided, based on RAPD analysis, into about five discrete groups—a southern group, a northeastern group, single population groups from the Marquesas and Hawaii groups, and, lastly, a divergent group from Rennel Island. They indicated that differentiation of coconuts in the Pacific has taken place on a geographical basis.

Thirdly, it is fairly well documented that beginning about the mid-19th century, all the European colonial powers—the French, British, German, Dutch, and the Portuguese—began setting up extensive coconut plantations in their respective colonies in east and west Africa, the Caribbean and Atlantic coasts, South and Southeast Asia, New Guinea, and the west Pacific islands. This continued for almost the next 100 years till the early years of the Second World War.

This process would have naturally led to a neglect of the local populations, which in turn could have dearly caused loss of variability in the aforementioned coconut-growing regions. These factors would have also

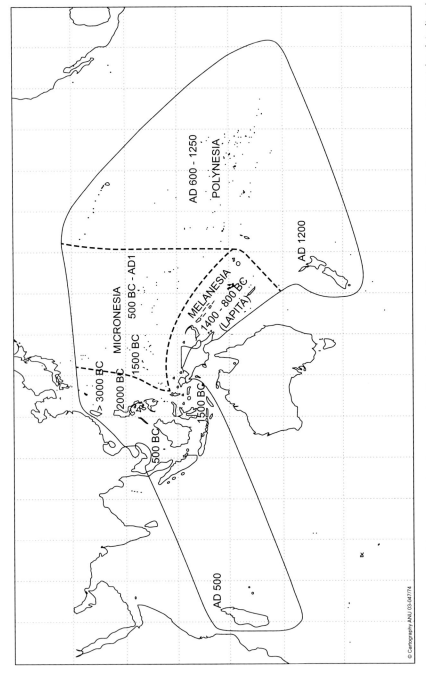

Figure 7.4 Distribution of the Austronesian language family in the Indian and Pacific Oceans, from Madagascar to Easter Island, indicating arrival timings. These dates also track the initial human settlement of Remote Oceania (as delineated in Fig. 7.1). *Reproduced from Bellwood, P. (2004). First farmers: The origins of agricultural societies. Malden, MA, USA: Blackwell Publishing. 360 pp. with permission.*

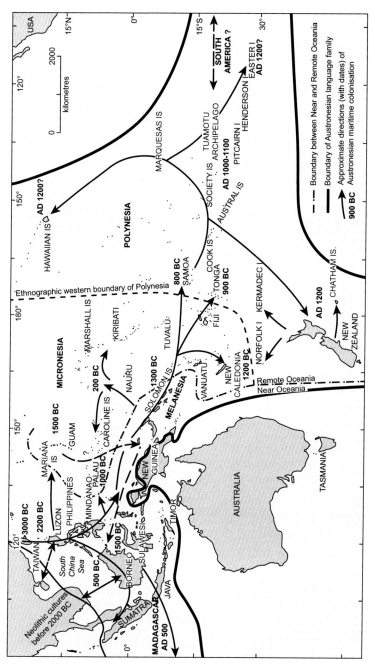

Figure 7.5 Holocene population movements through insular Southeast Asia and across Oceania, according to archeological and comparative linguistic data. This map is also published in Bellwood (2013). Original by Multimedia Services, Australian National University. Pacific Ocean showing traditional cultural regions and new secular divisions in to Near and Remote Oceania. *Reproduced from Bellwood, P. (2013). First migrants: Ancient migrations in global perspective (308 pp.). Malden, MA, USA: Wiley Blackwell with permission.*

led steadily to a miscegenation of remaining variability and, consequently, loss of native variability. This is apparent in the current state of coconut diversity, especially because the coconut is predominantly a cross-fertilizing species.

2.7 Harries' Views

Harries (1978) published a landmark paper on the evolution, dissemination, and classification of the coconut. Regarding the origin of the coconut, Harries proposed "that a continuous cycle of natural selection could produce a coconut palm with the following characteristics: perennial growth (50–100 years), few fruits (50–100) per year, large fruit size (1–2 kg), thick husk (up to 70% fresh weight), much endosperm (200–300 g), slow germination (more than 200 days)." He had not given the specifications and nature of the primitive coconuts—his "primordial *Cocos*."

Harries, Baudouin, and Cardena (2004) observed that when the European explorers discovered the uninhabited islands in the Pacific Ocean, they suggested that such primordial *Cocos* would also have occurred in Southeast Asia. "If the first cultivators in that region [where crops such as rice, taro, breadfruit, bananas, and sugar (sic) were domesticated] applied even simple selection to this already acceptable plant, then any change under cultivation would be detrimental to natural dissemination by floating," Harries added.

Harries felt that, contrary to the general consensus favoring human-assisted dissemination over natural dissemination, "both (human-assisted and natural) have been equally important, but at different times… "The primordial coconut might have originated in the continents which emerged or the land which submerged when Gondwana land divided, became established on atolls or regions where plant and animal competition had been destroyed by volcanic activity. This allowed the evolutionary cycle described above to take its course. The coast of America is just at or beyond the limits of natural dispersal. Africa was close or within the limit, but large stretches of coast were unfavourable, either because of (predation by) animals or because it was too dry for all or part of the year. Australia was also too dry…. Asia alone provided a suitable climate." The author did not indicate either the time or the context in which the two "different times" were employed for dissemination by the species. Incidentally, several of the previous observations have been since found to be at variance with the presently known facts or positions.

The author appears to have overlooked the fact that when Gondwanaland divided, neither the genus *Cocos*, and not even the subtribe Attaleinae to which it belongs, differentiated, nor were they present in this region, as per the available phylogenetic records (cf. Chapter 5; Table 8.8). The author then observed that "coconut grows in mainland southeast Asia not because man took it there, but because he was there to receive it when it arrived … The coconut not only became established, it was subjected to selection under domestication. A different coconut variety emerged from this process, and it was this variety that became dependent on, and widely disseminated by man." It is unclear if, by this statement, the author implied that the coconut evolved in continental Southeast Asia. He did not give any temporal framework for this process.

Mahabale (1978) discussed the origin of coconut mainly from a paleo-botanical angle. He concurred with Beccari (1917)'s view that the coconut arose from the southeast African species, *Jubaeopsis caffra*. And, this happened in some islands off the Indian archipelago including Sri Lanka, Nicobar, or some extinct island in the Indian Ocean. The author did not state how he arrived at this conclusion.

While discussing the biogeography of coconuts, Harries (1992) mentioned the natural occurrence of coconuts in some regions of the world. For instance, Buckley and Harries (1984) reported the occurrence of self-sown coconuts on a Great Barrier Reef island, off eastern Australia, and the presence of "wild types" of coconuts in eastern Philippines (Gruezo & Harries, 1984). This latter location was close to the place (Samar, Central Philippines, 11–13°N, 124–126°E approximately), where wild coconuts had been first discovered in the 17th century by the Spaniards (Alzina in 1668, according to Greuzo & Harries, 1984). Greuzo actually collected the material and confirmed its identity as the niu kafa type. The wild palms were growing in a mangrove swamp. The authors did not give any more details. The "Palaeo-"studies summarized in Chapter 4 have shown that coconut was part of the naturally occurring vegetation in much of Southeast Asia and the west and central Pacific long before the modern present-day humans arrived there.

Harries (1995) stated the following in the revised edition of Evolution of Crop Plants. This is being given in more detail because of the importance of the volume and the several nonconventional and speculative observations contained in it.

"Today, the coconut is an oil crop. Yet for natural selection and human domestication (sic), the water, shell, and husk were more important than the oil content… The water was more important than the cavity during the period of domestication. Selection over many generations increased water content at the expense of husk thickness… This was to the detriment of natural dispersal. More water in the immature nut means a larger cavity in the ripe nut and increased buoyancy. However, viability when floating decreases (sic) because the thinner husk gives less protection and a faster rate of germination. Eventually, floating became unimportant, as human activity accounted for further dissemination. Coconut then went inland or upland and to regions beyond the range of floating. Although human selection was mainly for water content, palms with dwarf habit or striking fruit colour also found flavour[sic]. All other parts of the palm and the other components of the ripe fruit became useful under domestication."

Harries (1995) observed further that "fossils and archaeobotanical remains support ethnobotanical arguments for a southwest Pacific origin. The coconut evolved to float between the coasts and offshore islands of the Palaeolithic Tethys Sea… As the tectonic movements caused the Tethys Sea to disappear, the primordial coconut emerged into the developing Indian and Pacific Oceans."

Incidentally, most of the previous observations are at variance with our present-day knowledge in the areas of geology, archeology, and phylogeny.

Harries (1990) then concluded that: "the most likely region for coconut domestication is (was) Malesia, on the coasts and islands between southeast Asia and the western Pacific." Although both the wild and domesticated types are suitable for commercial production, they contrast in many ways—in the germination rate and plant habit, flowering pattern, fruit form, and proportions of fruit components.

2.8 The Coconut in Australia

There have been some reports about the occurrence of the coconut in the northern regions of Australia from the latter half of the 19th century (Bailey, 1902; Mueller, 1867; Thozet, 1869). Buckley and Harries (1981) observed that coconuts in Australia were generally presumed present in only operating or abandoned plantations, or as small naturalized populations. In the present study, the senior author (Buckley) surveyed strandline sites, and found coconut plants in two locations, Lizard Island, north Queensland (7 km² area, 14°40′S, 145°30′E), about 30 km northeast of Cape Flattery. Both the populations were found within the range of storm tides. In one location, one palm was c. 10-m tall, and in the other, they were shorter. All of them bore abundant fruits of the long, angular thick-husked niu kafa type. They were uncertain if they were self-sown or not. They concluded "that in the sense of having reached Australia unaided by man, the coconut is therefore indigenous to Australia."

Dowe and Smith (2002) have observed that presently coconut palms form a dominant part of the coastal vegetation throughout much of tropical Australia. Incidentally, in Australia, coconuts in the form of nuts were first sighted in 1770 during the voyage of the *Endeavour* along the east coast of Australia (Beaglehole, 1962).

There is no need to assume that the presence of coconuts on the continent is the result of drift coconuts. This is because as recently as 8000 Ya, Australia and New Guinea constituted one landform known as Sahul (Fig. 7.3). And coconut has been known to be home to natural stands of coconuts from prehistoric times.

3. ORIGIN OF THE PACIFIC OCEAN

The Pacific and Indian Oceans have been the central theater in the evolution of the coconut.

The origins of what now constitutes the Pacific Ocean extend back to 750 Mya to the rifting of the Proterozoic continent, Rodinia. It split into two parts creating the Panthalassic Ocean, the ancient predecessor to the Pacific Ocean. By 400 Mya, the supercontinent Pangea began to form. This separated the Paleo-Tethys Ocean to the east and Panthalassic Ocean to the west. By 167 Mya, the sea floor spread to the west of Pangaea. Incidentally, this seafloor is now identified as the oldest ocean floor of the Pacific plate. By 140 Mya, Gondwana (the southern portion of Pangaea) had begun to fragment leading to India separating from Madagascar to form the Indian Ocean. The Tethys Ocean then lay between eastern Gondwana from eastern Laurasia (www.scotese.com).

About 50 Mya, the Earth was beginning to attain its current form with the oceans and continents that we have today. By then, India had begun to collide with Asia closing the Tethys Ocean and beginning the formation of the Himalayas and the Tibetan plateau. At c. 35.5 Mya, Australia finally separated from Antarctica and moved northward to its present position. During this time, the Pacific plate expanded to occupy finally two-thirds of the current Pacific Ocean (www.scotese.com). Paleomagnetic stratigraphy and deep-sea drill cores support this model. Currently, there are three plates in the Pacific Ocean, all originating from an initial one plate.

The large number of islands in the Pacific Ocean are assumed to have been formed by one of the following five major processes (Neal & Trevick, 2008).

1. Formation of linear volcanic island chains and seamounts on the above plates, either by mantle plume or propagating fracture origin, possibly synchronously c. 8 Mya.

2. As volcanic islands move away from their birth place, the oceanic crust become progressively colder and denser, and gradually subsiding. In tropical waters, coral growth forms fringing reefs, and its further subsidence leads to atoll formation.

3. By flexing of the lithospheric plate leading to uplift. At scattered locations in the South Pacific, coral reefs that once grew above an extinct volcano have been uplifted above sea level. Such islands are composed entirely of limestone, e.g., Niue (height 65 m), Henderson Island (height: 30 m), etc. Rennel Island in the Solomon Archipelago (height: 120–150 m) with 660 km^2 area is the second largest raised coral island in the world.

4. Fragments of continental crust rotating away from nearby land masses to occupy isolated positions, e.g., Zealandia displaced eastward from Gondwana 83–54 Mya by mid-ocean spreading of the Tasman Sea, but now largely submerged—New Zealand, Chatham Islands, New Caledonia, etc. Their geological histories may extend back 40 Mya.

5. Formation of island arcs on the Pacific margins mostly due to subduction of the Pacific plate. They are seen in northern and western Pacific; eg, Aleutian (35–55 Mya), Japan–Kuril (since at least the Cretaceous), Tonga–Kermadec subduction zone (c. 45 Mya).

In addition to the previous five processes, several other independent but minor processes may result in island formation, e.g., globally fluctuating sea-level changes during the Quaternary period.

Usually, glacial phases were more protracted (41–100 Ky) than interglacials (c. 20 Ky). The sea levels have been estimated to change 60–90 m (Clark et al., 2009, 2012) during this period—obviously, lower levels during the glacial phases and higher levels during the interglacial period, as now. A dominant feature of the Pacific islands is that many of them appear in chains. Now, 11 such linear chains and 10 exemplar archipelagoes (Japan, Izu–Bonin, Palau, Solomons, Fiji, New Caledonia, New Zealand, Society, Galapagos, and Hawaii) are identified in the Pacific Ocean (Neal & Trevick, 2008).

4. ATOLL ORIGINS AND ECOLOGY

4.1 Introduction

At this point, it would be instructive to get an overview of the nature, properties, and occurrence of atolls or coral atolls. This account has been taken primarily from Wiens (1962) and Wikipedia (2015, modified on 17 May 2015). The former one is considered as one of the most authoritative publications on atolls origins and ecology (cf. Kirch, 2000).

A typical atoll is an annular reef enclosing a lagoon in which there are no promontories other than reefs and islets composed of detritus, or, it may be a "ring-shaped ribbon reef enclosing a lagoon."

Most of the atolls are in the Pacific and Indian Oceans. The Atlantic Ocean has only eight atolls. They are found east of Nicaragua in the Caribbean. The coral of the atoll often sits atop the rim of an extinct seamount or volcano that had eroded or subsided partially beneath the water. The lagoon forms over the volcanic crater, whereas the higher rim remains above water or at shallow depths. This permits the corals to grow and form reefs. They thrive only in warm tropical and subtropical waters of the seas. Hence, atolls are found in only the tropics and subtropics.

There are only about 25 large atolls in the world. Of these, six are submerged in the sea. In most atolls, the land area is very small in comparison to its total area. Atoll islands are generally low-lying, less than 5 m in height. In terms of land area, the largest atoll is Kiritimati (Kiribati, Pacific Ocean; 1°58′N, 157°27′W) with 321–575 km^2 land, 160 km^2 main lagoon and 168–319 km^2 other lagoon areas. Lihou (1146 km^2 area; east of Queensland, Australia 17°25′S, 151°40′E) is the largest coral island in the world. It is followed by Rennell Island (660 km^2 area, Solomon Islands, West Pacific Ocean, 11°S 160°E). Atolls are the product of the growth of tropical marine organisms.

4.2 Atoll Origins

The following account has been taken from Wiens (1962), Atoll Environment and Ecology. Most of the Pacific basin is an area of great stability. It has been so for a long geological time. A subsidence of the Pacific Ocean basin of several hundreds of fathoms (one fathom = 1.8 m, 6 ft) had taken place in mid-Pacific mountains during the Tertiary (66–5 Mya) and Pleistocene (2.6–0.01 Mya).

Charles Darwin (1889) had proposed that atolls develop by growth of reefs and subsiding volcanic islands. Most geologists have since accepted Darwin's main premises about their origins.

Three drill cores made by the US Geological Survey in the 1950s in Funefutic Atoll, 2556 ft (774.5 m) in Bekim Atoll, and 4222/4610 ft (1279.4/1397 m) in Eneiretok Atoll showed a thickness of coral and algal lime structures down to a depth of 1100 ft (333.3 m). Several other studies have revealed that atolls are built up over geologic time, on relatively stable foundations when sea levels were several hundred meters lower than now, or that they were built on subsiding foundations that had sunk hundreds of meters since reef construction began. The growth of the reef usually keeps pace with the rate of rising waters. Normally, the average annual growth is about 2.5 cm.

Regarding the presence of coconuts on the atolls, Kirch (2000) had observed that atolls are a "most peculiar kind of island—and often, a precarious environment as well. An atoll requires critical adaptations, if human existence is to be sustained. The greatest single challenge facing life on an atoll is obtaining water, both for personal use and for growing crops. The dominant vegetation consists of coconut (drinking fluid, nourishing meat, and leaves for plaiting and weaving) and *Pandanus* (food and leaves); also present are breadfruit (*Arthocarpus altilis*), and swamp taro (*Cyrtosperma*)" (Kirch, 2000). It is difficult to assume that a coral atoll can be considered a secure ecosystem with its own biome and with the coconut as its principal component. At the same time, coconut is known to be a pioneer in establishing itself on new nearby islands including atolls. They are widely present in most tropical coral atolls, and form the main component on several of them.

4.3 Atolls and the Coconut

Charles Darwin in 1842 and Beccari (1916) had opined that coral atolls had a significant role in "coconut development." Harries has even gone to the extent of declaring that it was in the coral atoll ecosystem that the coconut developed into its present domesticated condition "long before any human interfere became possible, it was the pantropical presence of the coral atoll ecosystem that gave the sea borne *Cocos nucifera* the particular qualities that made it an endemic pioneer" (Harries, 2015; Harries & Clements, 2014). The authors did not, however, spell out any particular qualities of such a coral system which would have made it an "endemic pioneer," and what actually constitutes an endemic pioneer.

Harries (2001) had observed earlier that "a wild type (natural ecotype) must have existed somewhere in the Palaeogene (66.0–33.9 Mya); and the natural ecosystem had a very marked effect on the ecotype of coconut which evolved in it and can still be found today" (Harries, 2001, 2015). He did not clarify this process further. The author then observed further that "in the evolutionary development of the coconut, 19th century plantations were probably not the first time that it became the dominant (native) species before more competitive (woodland) species arrived" (sic) (Harries, 2015). The author has not given any supporting evidence for making any of these observations. Harries and Clements (2014) had observed further that "*Cocos nucifera* originated and dispersed by populating emerging islands of the coral atoll system, where establishment conditions (sic) exercise high selection pressure for survival… Wild coconuts evolved naturally and dispersed by floating without any need for human assistance."

Several authorities (Bellwood, 2013; Kirch, 2000) have observed that in the Pacific Ocean coconuts had been dispersed primarily by the Polynesian voyagers. Further, and even more importantly, we have seen that naturally occurring coconuts were present as a component of the natural vegetation in at least much of the western and middle Pacific Ocean islands long before modern humans set foot there. In addition, in the time frame indicated by Harries (Palaeogene, 66.0–33.9 Mya), the tribe Cocoseae was only just beginning to diversify in South America (Baker & Couvreur, 2013b; Meerow et al., 2014).

5. PEOPLING POLYNESIA VIS-À-VIS THE FOOD PRODUCTION SYSTEM

Various authors who have researched in region (Bellwood, Fox, & Tryon, 1995; Gosden, 1992; Kirch, 2000, etc.) have observed that Polynesia, east of the Bismarck archipelago (Papua New Guinea, 2°30′S, 150°0′E) was resource poor having only a few edible plants and tubers. In all the tropical Pacific islands, generally, the coastal strand vegetation is remarkably similar. It is dominated by *Pandanus, Barringtonia asiatica*, and the "ubiquitous coconut" (Kirch, 2000). A form of coconut was present naturally as far as central Polynesia (Cook Islands) prior to human settlements. Later, the colonists appeared to have brought canoe loads of coconuts. Many of the coasts had been covered by mangroves (mainly, *Rhizopora*) and sago palms (*Metroxylon*) (Kirch, 2000).

5.1 Long Pause

Several authors have pointed out a long break in the eastward colonization (Fig. 7.5) of the Pacific after about 1000 BCE (Bellwood, 2013; Kirch, 2000). Although Tonga, Samoa, and other central Polynesian islands were colonized by 1000 BCE, the earliest archeological assemblages from eastern Polynesia (northern Cook Islands, Marquesas, etc.) date to around 200 BCE only. Unequivocal radiocarbon dates from eastern Polynesian assemblages are datable to only c. 600 CE and later periods.

Parkes (1997) presented evidence for the presence of coconut in Atiu Island (Cook Islands, 17°S, 160°W) by 7820 ± 70 BP. Kirch (2000) stated that this showed that the coconut was present in Remote Oceania well before human colonization. He added also that in a similar manner, the total absence of coconut pollen in the deeper levels of the cores from the Pago swamp and Palau indicated that wild coconuts were not present in Micronesia before human settlement.

Gosden (1992) had pointed out that a general gradient of poorer environments in Polynesia is evident as one moved from Bismarck Archipelago to Tonga, Samoa, etc.

When the first humans moved into the Pacific Islands c. 33,000Ya, they subsisted on a combination of the foraging of plant and marine foods, and some hunting on land, but more by harvesting from the sea. But by 3500 BP, they were raising all the major animals and plants for food. White yam and taro complex (aroids: taro, swamp taro) supplied the basic staples, and they were being supplemented by breadfruit (*Artocarpus attilis*), *Canarium* spp., and sago to varying intensities, along with pig, dog, and chicken. Incidentally, Gosden had not included the coconut as one of the plants in the list. Although there is evidence for a long period of exploitation of natural resources from the terminal Pleistocene, i.e., by gathering and hunting, there was a clear break with the earlier systems with the arrival of the Lapita people cultural complex (1350–900 BCE). Then regular use of seafaring and social skills came along for colonizing of the Pacific. They also introduced diversity in food production systems in the western Pacific (Gosden, 1992).

The earliest initiation of Lapita culture may be datable to 3440 ± 100 BP (uncalibrated) (Kirch, 1997). "Between 1350 and 900 BC, Neolithic colonists in Island Melanesia (excluding New Guinea) and western Polynesia (Tonga, Samoa, and nearly[sic] islands) left an extremely clearcut[sic] trail of sites belonging to the Lapita cultural complex" (Bellwood, 2013).

5.2 Colonization of East Polynesia

The 15 archipelagos of east Polynesia—including New Zealand, Hawaii, and Rapa Nui—were the last habitable places colonized by the humans. But, there have been problems in resolving the timing and pattern of colonization.

Wilmhurst, Hunt, Lipo, and Anderson (2011) conducted a metaanalysis using 1434 radiocarbon dates. They found that colonization occurred in two distinct phases: earliest, in the Society Islands during ~1025–1120 CE, four centuries later than previously assumed, then, after 70–265 years, dispersal continued in one major pulse to all remaining islands during ~1190–1290 CE. They concluded that this would explain the remarkable uniformity of culture, human biology, and language in east Polynesia.

5.3 Food Production Economy of the Pacific

Kirch (2000), in his monograph on Pacific prehistory and archeology, On the Road of the Wings, has dwelt at some length on the food production system of the region. The proto-Austranesians had their own vocabulary for the locally important crops and animals as taro, bananas, breadfruit, sweet potato, coconuts, pigs, chicken, and dog, and also for canoes and activities associated with seafaring. This has been confirmed also by well-known linguists like Blust and Pawley (also, Bellwood, 2013; Bellwood et al., 1995; Yen & Mummery, 1990).

Lapita people were the first modern humans (*Homo sapiens*) to occupy most of the western and central Pacific region (Remote Oceania and western Near Oceania). They were inveterate sailors. Wherever they sailed in Remote Oceania, they went fully equipped for establishing permanent settlements. They are believed to have propagated 28 plant species. Movement of 15 of them have been attested by archeobotanical remains, and the rest by linguistic and other inferred evidences (Gosden, 1995; Hather, 1992). This includes also the coconut.

Direct evidence for human settlement of the Marianas by 1500 BCE has been confirmed from the composition of sediment cores dating them to beginning 7900 BCE from Orate Peninsula and Pago River Valley (Guam, 13°27′N, 144°45′E). However, records of human occupation, as evidenced from charcoal, have been obtained from only 3550 BCE. Incidentally, coconut pollen dating to this period has also been recovered. This indicates that coconut was either an introduced species at this period, or more possibly, a component of the naturally occurring vegetation, in the region.

Ward, Athens, and Holton (1998) collected 22 sediment cores from Babeldaob Island (Palau, Belau (7°30′N, 134°34′E)) and Western Caroline Islands (Micronosia, 0°N, 150′E, just east of the Philippines). They analyzed three of them for pollen flora. They showed a sudden increase in savanna grasslands by 3100 cal BP, along with substantial increases in carbon, suggesting anthropogenic changes. Pollen of two economically important plants, betelnut (*Areca catechu*) and coconut (*Cocos nucifera*), first appeared as early as 4229 cal BP. The author took it to mean the presence of humans in

Palau by at least 3000 cal BP, and possibly, by 4200 cal BP. However, it is more plausible that the presence of the arecanut and coconut was a consequence of these two species occurring in these islands as part of the natural vegetation. The presence of *A. catechu* in this region has been confirmed as a component of the local vegetation (Dransfield et al., 2008).

Early Polynesian communities were supported by much the same horticultural and maritime economy as their Lapita ancestors, cultivating tuber crops including taro and yams, and tree crops such as the coconut, breadfruit, bananas, and Tahitian chestnut (*Inocarpus fagifer, Leguminosae*). Good lexical evidence for the practice of shifting cultivation is also available. They appeared to have kept pigs, dogs, and chickens, but only chickens have been represented in archeological deposits. Fish and shellfish provided most of the protein (Yen, 1990). Linguistic reconstructions have indicated these people were using a wide range of fishing methods and using more than 150 different kinds of fish (Clark, 1991; Hoaper, 1994).

Recently, Kinaston et al. (2014) studied of Lapita diet using stable isotope remains of human bones. They used material from a site that had 68 inhumations—at Teouma, close to the present Port Vila, Efate Island, Vanuatu (17°44′S 168°17′E). Most of the food consisted of animal and marine items. Plant "remains" consisted mostly of bananas (*Musa* sp.) aroids, and yams. Some sites showed evidence of Arecaceae (in decreasing quantities) and Poaceae (in increasing quantities). They appeared to indicate vegetation clearance, likely associated with new human settlement and horticultural activities. They found coconut remains dating to c. 5000 BP on Ancityum Island.

6. NATURAL DISSEMINATION OF THE COCONUT

While discussing this topic, we should keep in view that the coconut was part of the naturally occurring vegetation from the western Pacific up to at least. Cook Islands (eastern Polynesia 17°00′S, 160°00′W)—possibly even as far east as the Society Islands—even before the first humans set foot there. Incidentally, the tropical and subtropical South American coast is only 5000–6500 km from this region.

The traditional assumption has been that the wide occurrence of the coconuts on the Pacific Ocean islands and atolls has been achieved by dissemination by ocean currents. Since then, several scientists familiar with the Pacific Ocean have shown that wide dissemination of coconuts has happened by dissemination by early seafarers. We now know that early Polynesians were inveterate seafarers and they used to carry coconuts with them primarily as a source of drinking water, and possibly also for propagation. Most smaller Polynesian islands do not have perennial water sources. Researchers agree, however, that dissemination through ocean currents could certainly have happened to nearby islands. At the same time, it has now been seen that the wide occurrence of coconuts has been because the coconut was a component of the natural vegetation in the Pacific Ocean islands. Please refer Chapter 6, Biogeography, for details.

It has been suggested by some authors that dissemination of the coconuts had taken place also by the establishment of seedlings derived from the surplus coconuts left behind by the seafarers when they were leaving the islands. There is, however, a glitch in this. If the coconuts had been taken primarily for water, then the seafarers would have ideally taken immature nuts (4–5 months old) only, because the cavities of such nuts will not only be filled with water but the "meat" will also be tastier. If, however, they had chosen to carry mature nuts (10–12 months old), then their water content would be low as only about one-third of the nut cavities would contain water. Further, the meat, though bland in taste, would have been found to be more nourishing and useful as food. Nuts younger than 10 months are unlikely to germinate and grow into normal healthy palms.

There are a few incontrovertible evidences for natural establishment of coconuts in recent times. A widely quoted one is the presence of coconuts on the Vertaten Islands, northwest of Krakatau in Sunda Straits (between Java and Sumatra) (Hill, 1929; Hill & van Leeuwen, 1927). The island was both uninhabited and uninhabitable. In April 1919, van Leeuwen found coconut palms growing in drift mud in the Island. In December 1927, volcanic activity began on the Krakatau Island, and a new island appeared nearby in January 1928. It was 275-m long and 3-m high. This island, however, disappeared in July 1929. Then, a third, Krakatau III, appeared on July 3, 1930, which also disappeared, on August 11, 1930. A fourth island, Krakatau IV, rose from the sea. It was about 1-km long, 40-m high in January 1932. The beach contained black soil, a great quantity of pumice, and wood logs washed ashore. Soon, many seeds of the *Barringtonia* association began to germinate. This included also 41 coconut seedlings. By November 1932, this island too had disappeared! Nevertheless, this proves the point that coconuts could be washed ashore on land without the aid of humans, and that they could germinate and establish themselves without human assistance.

Several other authors and naturalists, too, have recorded observing coconut palms on some uninhabited islands of the Indian and Pacific Oceans. This included Charles Darwin in 1837 in Cocos (12°10′S, 96°55′E) and Keeling

(12°10′S, 105°55′E) Islands, south and southwest of Java (Indonesia), Sauer (1967, 1971) in the Seychelles Archipelago, and so on.

There have been also numerous reports of sightings of the coconut on the Pacific coast of the Americas and some the nearby islands by some early travelers during the Pre-Columbian period. But, practically all of them have been since discounted, as we have noted already.

7. NEW GUINEA AS AN INDEPENDENT CENTER OF AGRICULTURAL ORIGINS

Hather (1992) possibly was one of the earliest authors to indicate that New Guinea represented an independent center of origin of cultivation. This included sugarcane and banana. He pointed out also that the majority of the preserved plant remains in the Pacific were recovered as charred remains which indicated the presence of humans. In Melanesia, humans subsisted on harvested nuts and fruits, as was earlier pointed out by Barrau (1961). Specifically, they subsisted on coconuts, bananas, breadfruit, sago palm, several *Canarium* and *Barringtonia* species, and *Inocarpus fagifer* (Polynesian/Tahiti chestnut). Most sites of early human occupation showed also an abundance of nut shell remains. Plant resources were poor in Polynesia (Yen, 1995).

7.1 Coconuts in Papua New Guinea

Dwyer (1938) commented on the then state of coconuts in New Guinea. Coconut was then the most important agricultural product of the country. Dwyer observed that most of the native palms of Matty Island (Bismarck Archipelago, Papua New Guinea; 2°55′S, 146°22′E) produced nuts in which the kernel formed only a small portion, and the amount of husk was large. The diameter of the husked nut was only c. 5.75 cm. The complete oval-shaped fruit measured c. 25×17 cm at the shoulders, the nut size was 10.0×5.5 cm, the husk thickness varied 8.25–5.50 cm, the central cavity had a 3.0 cm diameter, and meat, 1.3 cm thick. The husk was very difficult to open. The overall yield of palms was low. A good percentage of the palms belonged to this category.

It appears from the description that these palms are much more primitive than the niu kafa type, and a more primitive form than had been reported from east of the Philippines (Greuzo, 1990).

In the following chapter, Origins, we shall attempt a synthesis of all the information presented into developing an integrated account of the origins of the coconut.

CHAPTER 8

Origins

1. INTRODUCTION

In the previous chapter, we had reviewed the past literature on the origins of the coconut. Readers would have noticed two points: the general lack of agreement on any of the aspects of the origin of the coconut; the wide gap in information on most aspects. Hence, much experimental work will need to be done before we can make more definitive conclusions.

An enquiry into the origin of a crop plant involves investigating four aspects: place of origin, time of origin, ancestral species/taxon, and mode of differentiation from the wild to the cultivated form. We shall now take up these aspects in the following section.

2. PLACE OF ORIGIN

2.1 Past Propositions

In contemporary times, De Condolle (1886) was the first to make any suggestions on this topic. In the first edition of his classic, Origin of Cultivated Plants in 1882, De Condolle had favored the origin of the coconut in South America. But, in its second edition (1886), he suggested the Indian archipelago as the place of origin. In addition, he had given his reasons also for changing his opinion.

An archipelago is by definition an extensive group of islands. The most obvious example that comes to the mind is the Malayan archipelago—the group of islands lying to the south and southeast of continental Malaysia and constituting the present day Indonesia: Sumatra, Java, Borneo, Celebes, New Guinea, and the hundreds of small islands dotting their surroundings. In India, strictly, there is none like that to be reckoned as an archipelago. However, De Condolle might have meant the Lakshadweep (Laccadive) Islands—the group of c. 260 tiny to small coral islands lying to the southwest of peninsular India—and Maldives—a large group of 1000-plus small islands lying further south of the Laccadives, to the southwest of Sri Lanka and India in the Indian Ocean. Only eight islands of Lakshadweep and c. 40 islands of the Maldives are occupied by humans.

The Coconut
ISBN 978-0-12-809778-6
http://dx.doi.org/10.1016/B978-0-12-809778-6.00008-5

All the islands are only 1–3 m above mean sea level. All of them are coralline in origin. Most harbor natural stands of coconuts. It is unclear what impelled De Condolle to propose these islands as the center of the origin of coconut.

Cook (1901, 1910), a US botanist, was a very strong proponent of the origin of the coconut in South America. He gave accounts of numerous travelers of South America during the post-Columbian period about their finding coconut palm stands on the Pacific coast islands to support his contention. Cook had conceded that the noted palm scientist von Martius (1794–1868), and the well-known naturalist, A.R. Wallace (1823–1913), had both favored an Asiatic/Polynesian origin of the coconut. Nevertheless, Cook (1901) felt that, according to him, the probabilities favored the alkaline regions of the Andes of Colombia as the center of origin of the coconut. Heyerdahl (1950), the explorer and archeologist from Norway, firmly believed that the American Indians had carried the coconut from South America to Polynesia.

We have discussed already the subject of Pacific Ocean exchanges.

There are both strong proponents and opponents of these two divergent viewpoints. This debate was activated recently by the discovery of chicken bones from an archeological site in Peru that was dated to 1304–1424 CE (Storey et al., 2007). This is a pre-Columbian date, and the chicken has been accepted as a post-Columbian introduction to the New World.

Beccari (1917) was a forceful contestant of Cook's (1901, 1910) proposals. Beccari stated that the coconut might have originated in the islands lying in the eastern Indian Ocean, or in some other lands or islands existing in formal times between Africa and India. He proposed further that Sri Lanka and Cocos Islands (formerly, Keeling Island) lay almost in the region where the coconut assumed its present character. The only entity that could be considered to fill the description of Beccari is the Maldives archipelago (and by extension, the Laccadives also). Most islands in the Maldives are covered with natural stands of coconut. We do not have any more information on the coconut in the Maldives. However, the coconut is one of the staple food items of the Maldives islanders. It is used there in a wide variety of ways.

Mayuranathan (1938) felt that biological evidences indicated that the coconut evolved in the southwest region of New Guinea that is now submerged in the sea. However, he did not give any reasons for indicating this region. Child (1974) expressed helplessness in "assigning a precise habitat" in the Pacific Ocean between 80°E and 180°E, i.e., approximately between Sri Lanka in the west and Fiji in the east. In his last publication, Purseglove (1985) stated merely that the domestication of the coconut took place in the Pacific Ocean.

Harries (1978) had not indicated any specific location or region as a likely center of the origin of the coconut. Nevertheless, in his subsequent publications, he suggested two regions covering broadly Southeast Asia, the Philippines, and the western and southern Pacific Ocean—the Malesian region. Then, Harries (1981a, 1981b) observed that coconut arose in "isolated coral systems." In Harris (1990), he indicated the southwest Pacific as the region of origin. Geographically, the southwest Pacific Ocean covers the sea that is north of New Zealand, east of Australia, and south of New Guinea. The region includes New Caledonia (21°S, 165°E), Vanuatu (15°S, 168°E), and the Solomon Islands. In a later paper, Harries (1990) specified the region of domestication as Malesia, "in the coasts and islands between southeast Asia and the western Pacific." Malesia is a vast region covering the Malay Peninsula, Sumatra, Java, Borneo, the Philippines, Sulawesi, Papuasia (Bismarck Archipelago and Solomon Islands), and New Guinea (Dransfield et al., 2008).

Recently, Harries and Clements (2014) brought out another perspective suggesting that long-distance dispersal of the coconut palm took place by migration within the coral ecosystem. They observed that *Cocos nucifera* originated and dispersed by populating emerging islands of the coral ecosystem … Long-range dispersal is secondary, because spontaneous independent migration of coral polyps on a prolonged geological time-scale generated new atolls in new areas, where coconuts would be among the earlier inhabitants. The coconut plant became an itinerant pioneer endemic and on suitable beaches on volcanic or large islands and continental coastlines. We have covered this aspect already.

We have reviewed in the last chapter the various aspects of coral geology and ecology. Atolls are basically inhospitable for life systems, and consequently, an improbable habitat for a biome to develop on them. In atoll island systems like the Maldives, Lakshadweep, etc. where scores or even hundreds of islands of very small size occur, dispersal and establishment of coconuts on nearby islands is possible. This process may be continuing even now also. It may also have been possible for the coconut to disperse from one volcanic island to another one nearby, and establish there, as per the observations of various established Pacific Ocean scientists. Coconuts are not known to have been a competitive colonizer, but have been a component of the strand vegetation. At the same time, many authors consider ocean biota as a distinct ocean biome.

We shall now look at the available evidences to find the answer to the place of origin.

2.2 Available Evidence

2.2.1 Fossil Evidence

Readers are advised to refer Chapter 4 in which the earlier work on the topic has been reviewed. The results are tabulated in Table 8.1.

Cocos-like fossils have been recovered from 15–20 locations (Table 8.1): about 10–12 from New Zealand [equal numbers from North Island (from near Auckland) and South Island (from near Canterbury); another half dozen or more from central India—from the Deccan intertrappean zones—and one each from Queensland (Australia) and northeast Colombia (South America)]. All of them are from the Tertiary period, but from different epochs. Nevertheless, this is a long time span (65–2.8 Mya). The material from New Zealand has 3.5–5.0-cm long fruits and those from central India, 11–13-cm long fruits, except for Prasad, Khara, and Singh (2013), etc. There is also the report of a silicified stem of an entire *Cocos*-like stem of *Palmoxylon sundaram* from the intertrappean beds (Sahni, 1946). Its anatomical studies had shown its close resemblance to that of the coconut, according to the author. That from the Chinchilla Sand Formation (Queensland) had 10.0-cm long fruits, and from Colombia 25.0-cm long fruit, but the sample was only 4–8-mm thick. The photograph (Fig. 4.6), however, shows a close likeness to coconut.

About the fossils from North Island, New Zealand, Hayward, Moore, and Gibson (1960, and earlier papers) had suggested that they were possibly fruits of *Parajubaea coccoides*. The fruits from central India and the Chinchilla Sand Formation (Australia) come within the size range of primitive or mini or micro forms of cultivated coconut. Further, in their surveys of naturally occurring coconuts in the Nicobar Islands (8°0′N 93°30′E; Bay of Bengal) (Table 8.4) and Lakshadweep Islands (0°0′N 72°0′E; Arabian Sea; Table 8.5), coconuts of the size obtained from New Zealand (3.5–5.00 cm) have been obtained. This leads us to suggest that the fossil nuts recovered from New Zealand could also have been of a primitive form of the coconut, if not of one of the wild Attaleinae taxa. Further physical examination of the material and relating it will be needed to confirm this.

Table 8.1 Fossil records of *Cocos*-like fruits, leaves, and impressions

Sl. no.	Author (year)	Material	Location	Period
1	2	3	4	5
1	Berry (1914)	Earliest fossil palm: Costapalmate	Mideastern USA	Cretaceous (144–66 Mya)
2	Crabtree (1987)	Earliest pinnate leaves	Montana, USA	Cretaceous (144–66 Mya), Upper
3	Prasad et al. (2013)	Leaf and fruit impressions (mesocarp), $1.8 \times 1.5 \times 1.5$ cm	Mohgaon Kalan, Chhindwara, Madhya Pradesh (MP)	Tertiary: Maastrichtian–Danian (70–65 Mya)
4	Srivastava and Srivastava (2014)	*Cocos*-like fruits	Binori Reserve forest, Seom, MP, India	Maastrichtian–Danian (70–65 Mya)
5	Gomez-Navarro et al. (2009)	*Cocos*-like fossils of 5 species, 12 specimens; *Cocos*-like, 25.0×15.0 cm	Northeast Colombia, South America, open–cast mine	Paleocene (66–57 Mya) middle to late
6	Tripathi, Mishra, and Sharma (1999)	Petrified fruit, $13 \times 10 \times 6$ cm	Amarkartak, MP, India	Tertiary (66–65 Mya)
7	Shukla, Mehrotra, and Guleria (2012)	Petrified fruit, $13 \times 10 \times 6$ cm	Kapurdi, Barmer, Rajasthan, India	Early Eocene (57–36 Mya)
8	Ballance et al. (1981) and Campbell, Fordyce, Grebneff, and Maxwell (1991)	Nuts, 3.5–4.5 cm diameter	Hawkes Bay, north Otago, New Zealand; Canterbury, NewZealand	Eocene, Oligocene, Miocene(57–23 Mya)
9	Ballance et al. (1981), Campbell et al. (1991)	Scattered occurrence; fruits 5 cm long	Northland, New Zealand	Eocene, Oligocene, Miocene (55–7 Mya)
10	Hayward et al. (1960)	Nuts, 5 cm long	Waikiekie quarry, Bryndervyns quarry, N Island, New Zealand	Oligocene (33–37 Mya)
11	Berry (1926)	Nuts, 3.5×1.3–2.5 cm; fruits only slightly bigger	Manganui, north Auckland, New Zealand	Miocene (23–5 Mya)
12	Black (1996)	Nuts	Central New Zealand	Miocene (23–5 Mya)
13	Rigby (1995)	Premineralized *Cocos* fruits, 10.0×9.5 cm	Chinchilla, Qld, Australia	Pliocene (5.0–3.5 Mya)

2.2.2 Palynological Evidence

Readers are advised to refer to Chapter 4 in which the earlier work on the topic has been reviewed. The results are also tabulated in Table 8.2.

Parkes and Fenley (1990) recovered *Cocos* pollens dated to 7800Ya and 8000Ya from Cook Islands (Table 8.2). Because present-day humans (*Homo sapiens*) were not known to have reached this area by then, we can take that the coconut in a wild or primitive state, or some other taxon closely related to it, was present in that region as part of its natural vegetation.

Table 8.2 Archeological and palynological records of the coconut

	Author (year)	Location	Time period	Material
1	Fosberg and Corwin (1958)	Pagan, Mariana Islands, Micronesia	Late Quaternary	Seedling leaf
2	McCormack (2005)	Atiu & Mangia, Cook Islands	8600 BP	Pollen
3	Parkes and Fenley(1990)	Te Roto,CookA Islands	8000 BP	Pollen
4	Parkes (1997)	Atiu island, Cook Islands	7820 ± 70 BP	Pollen
5	Maloney (1993)	Several sites in Southeast Asia	6000 BP	Pollen
6	Spriggs (1984)	Ancilum Island, Vanuatu	5040 ± 70 BP 5420 ± 90 BP	Roots, numerous endocarp fragments
7	Hossfeld (1965)	Aitape, northeast Papua New Guinea (PNG)	4555 ± 80 BP	Nuts
8	Ward, Athens, and Holton (1998)	Palau Island, Micronesia	4229 cal BP	Pollen
9	Mathews and Gosden (1997)	Arawe Island, PNG	3840 ± 60 BP to 2900 ± 80 BP	Endocarp fragments
10	Kirch and Yen (1982)	Tikopia, south-east Solomon Islands	3360 ± 130 BP to 2695 ± 90 BP	Charcoal & nut remains
11	Parkes and Fenley(1990)	Lake Temae, Cook Islands	1500 BP	Pollen
12	Lepofsky et al. (1992)	Moorea, Society Islands	1270 ± 60 BP to 1360 ± 70 BP	Whole coconuts

Maloney (1993) reviewed the paleopalynological studies on the coconut. The presence of *Cocos*-like pollen was not obtained continuously from anywhere from the Miocene (23–5 Ma) to the present time. Based on palynological information, he indicated that the coconut was present in Southeast Asia from 6000 BP.

2.2.3 Archeological Evidence

The past work on this aspect has been reviewed in Chapter 4. The results are tabulated in Table 8.2.

Morcotte–Rios and Bernal (2001), who had reviewed palm archeological studies of South America, did not report the presence of any *Cocos* material from any of the New World sites. However, they had reported recovering material of several other members of the *Cocos* Alliance/subtribe Attaleinae (to which the genus *Cocos* belongs), such as *Attalea* s.l., *Astrocaryum* (now, *Syagrus*), *Bactris*, *Syagrus*, and *Elaeis*. Kirch (2015), who has conducted numerous archeological expeditions in the Pacific Ocean, has informed in a private communication that shell pieces of the coconut were being commonly found in several island archeological sites, but they had not been documented, mainly because they used to be very frequent and numerous. They were obviously the remains of the nuts left behind by the Polynesian voyagers after use.

There are still a few reports of obtaining coconut archeological remains from the Pacific Ocean (Chapter 4; Table 8.2) (1) Hossfeld (1965) from Aitape, Sepik district, northeast Papua New Guinea, dated to 4555 ± 80 BP in the context of human burial; (2) Kirch and Yen (1982), charcoal and coconut remains from Tikopia, Solomon Islands, dated to 3360 ± 130 BP and 2695 ± 90 BP, obtained in the context of strand vegetation. (3) Spriggs (1984), from a swamp site from Ancilum island, Vanuatu (Bismarck Archipelago) numerous pieces of coconut roots from the bottommost layer #6 (5040 ± 370 BP) and endocarp from layers #5 and #4 (5420 ± 90 BP and 5410 ± 100 BP, respectively). Thus, this is incidentally the oldest archeological remains of the coconut recovered anywhere. Incidentally, Bismarck Archipelago did not have any human occupation before 3500–2500 BP. (4) Lepofsky, Harries, and Kellum (1992), of early coconuts from Mo'orea, Society Islands dated to 1270 ± 60 BP and 1360 ± 60 BP. Incidentally, the Society Islands were first occupied by humans only by c. 650 BP (Bellwood, 2013). This indicates that the remains were of naturally occurring coconuts.

All the previous reports indicate that the coconut was part of the natural vegetation in this region long before present-day humans arrived there. The nuts possessed relatively thicker husk, oblong shape, and moderately thick shell, as some authors have commented, indicative of primitiveness.

2.2.4 Natural Occurrence of Wild/Primitive Coconuts

There are four recent reports about the occurrence of naturally growing coconuts. They are: (1) Dwyer (1938) from Matty Island, Papua New Guinea; (2) Greuzo (1990) from Eastern Samar, The Philippines; (3) Balakrishnan and Nair (1979) from the Nicobar Islands; and (4) Jerard, Rajesh, Thomas, Niral, and Samsudeen (2016) from the Lakshadweep (Laccadive) Islands. The latter two are from India, in the Indian Ocean.

It is also well known that naturally occurring coconuts are widely present in the thousands of islands of both the Indian and Pacific Oceans. The country of Maldives is an excellent example. It consists of an archipelago of c. 1190 islands in 20 atoll islands. Only about 200 islands are occupied by humans. Nevertheless, practically all the islands (except very small islands) contain naturally occurring stands of coconuts.

2.2.4.1 Matty Island, Papua New Guinea

Dwyer (1938) found in Matty Island, Papua New Guinea, wild forms of coconut. The fruit measured 25 × 17 cm in size, the nut, 10.0 × 5.5 cm, in the central cavity, 3.0 cm in diameter. The husk was very difficult to open.

Possibly, the micro and nana forms described in the literature may be analogous to these primitive forms. For instance, the fruit size of Laccadive Micro is 19.9 × 11.2 cm and that of Mini–Micro c. 6 cm only (Table 8.5). Watt (1892) mentioned a form from Sri Lanka "about the size of turkey egg." This means c. 6 cm. Watt observed that this form was "rare, but much appreciated." Many readers would have come across such "nana" forms from other regions of Asia as well.

2.2.4.2 Eastern Philippines

Greuzo and Harries (1984) reported the presence of primitive forms of coconut from the eastern Philippines. The senior author, Greuzo, on basis of an unpublished thesis written in Spanish by one F.L. Alzina in 1668 (Uichanco, 1931) visited a remote area in eastern Philippines—Guian peninsula, Eastern Samar province, facing Matarinao Bay (12°0′N, 125°0′E)— and recovered niu kafa-type wild coconuts. Subsequently, Greuzo (1990) explored two more provinces in the eastern Philippines, Eastern Samar and Surigao del Norte. He studied eight populations and conducted fruit component analysis of the fruits (nuts) of 124 palms. He measured four parameters—fruit weight, nut weight, shell + meat weight, and meat weight. He prepared scatter diagrams, but did not attempt to analyze them. He found two forms, talls and dwarfs, and, in both, two "subtypes" were recovered, consisting of large-fruited and small-fruited forms. Both forms had generally more than 55% husk weight. He stated that they conformed to the niu kafa forms. He reported also that some populations showed varying levels of domestication characters.

From the habitat and site description, it is apparent that the collections were made from areas where there were human habitation and activities. Naturally occurring and cultivated coconuts appeared to have been present in adjacent plots/areas. The old timers had informed the author that these wild/semiwild forms were locally called "Palau palms." According to oral history, people from Palau Island (Micronesia, 7°30′N, 134°30′E; 800–1000 km to the east) used to come to this region in ancient times with the onset of the northeast monsoon (November to February) in small sailboats, and then, they used to return back only in the following year during June to September with the reversal of wind flow (southwest monsoon?). These people used to bring with them long-fruited nuts, both as source of water and food, and leave behind the nuts remaining after use, when they were returning home. Also, if one of them died during their stay there, they used to plant a nut over his/her burial place.

The Philippines is known to possess a considerable level of variability in the present-day cultivated crop. Santos (1983) has classified them into 8 dwarf and 15 tall forms.

From the descriptions given in Greuzo (1990) and from analyzing the data included in the study (Table 8.3A and B), the regions where the author surveyed, it is apparent that the "wild coconuts" studied by the author and cultivated forms occur together in adjacent plots in the region. Overall, the fruit size of the fruits of even the so-called niu kafa coconuts of the Philippines appears to be much bigger and less variable than what has been obtained from the other two locations in South Asia (Nicobar Islands, Laccadive Islands), where naturally occurring coconuts have been described. Further, primitive coconuts were found to be occurring in two discrete forms: those possessing big fruits and small ones. From India (Nicobar and Laccadive Islands) the authors have reported a full range of size, from 1600 to >8200 mL in volume (Fig. 3.2; Tables 8.5 and 8.6). Notwithstanding this, a preponderance of bigger-sized fruits is apparent

Table 8.3 Fruit analysis of eight naturally occurring coconut populations of the Philippines
(A) Range and average

	Locality, palm number	Fruit weight (g)	Husk weight (%)	Kernel weight (%)	Shell weight (%)
1	Homonhon island (12 palms studied)				
	Range	1516–2271	45.5–70.1	11.9–30.6	7.2–15.3
	Average	1811	57.6	20.5	10.9
2	Subiyan island (4 palms)				
	Range	1674–1927	45.6–64.7	15.4–22.1	10.5–13.6
	Average	1809	51.6	19.6	12.9
3	Naga: Quinapondan (8 palms)				
	Range	599–1323	55.0–85.3	8.9–22.1	9.4–20.8
	Mean	813.1	65.7	168	13.8
4	Quezon, Siargao (18 palms)				
	Range	570–2030	64.8–95.9	1.0–16.0	3.1–10.8
	Mean	1204	63.2	9.8	7.9
5	Bay antian, Dinaghat (8 palms)				
	Range	540–1620	61.7–77.9	8.7–16.0	7.1–11.1
	Average	839.1	69.6	13.3	8.7
6	Luna, Dinaghat (11 palms)				
	Range	470–1980	53.2–64.0	11.6–24.0	11.2–17.0
	Mean	1270	51.5	18.1	12.4
7	Baltazar, Dinaghat				
	Range	430–1540	46.4–88.9	2.7–23.8	4.3–23.3
	Mean	1201.7	75.0	8.8	7.9
8	Magsaysay, Surigao (8 palms)				
	Range	580–1240	54.3–90.3	4.0–20.2	5.6–17.2
	Mean	819	71.9	16.4	15.3

(B) The highest and lowest

	Locality, palm number	Fruit weight (g)	Husk weight (%)	Kernel weight (%)
1	**Homonhon island**			
	H-9. Highest fruit weight	2271	70.1	11.9
	H-5. Lowest fruit weight	1516	56.4	20.6
	H-9. Highest husk weight	2271	70.1	11.9
	H-4. Lowest husk weight	2022	45.5	22.1
	H-2A. Highest kernel weight	1740	50.5	24.1
	H-9. Highest kernel weight	2271	70.1	11.9
4	**Quezon, Siargao island**			
	Q-31. Highest fruit weight	2030	62.6	9.9
	Q-11. Lowest fruit weight	570	73.7	10.5
	Q-40. Highest husk weight	970	95.9	1.0
	Q-31. Lowest husk weight	2030	62.6	9.9
	Q-6. Highest kernel weight	1000	66.0	16.0
	Q-15. Lowest kernel weight	920	87.0	5.4
3	**Baltazar, Dinaghat**			
	B-163. Highest fruit weight	1540	84.4	3.9
	B-154. Lowest fruit weight	430	53.5	18.6
	B-174. Highest husk weight	1440	88.9	3.5
	B-168. Lowest husk weight	840	46.4	23.8
	B-168. Highest kernel weight	Do	Do	Do
	B-173. Lowest kernel weight	1480	87.8	2.7

Characteristics of above sites.
Homonhon island: Along the coast, c. 530 m from shoreline, few trees on hill grounds.
Quezon: Siargao island: Plantation c. 1 km from nearest shoreline.
Baltazar: Remnant of Nype-mangrove swamp forest, c. 150 m from the seashore.
(A and B) Calculated from Gruezo (1990).

in the Philippine material. This may indicate that the local people have been selecting bigger-sized fruits for planting in the region. The husk content of most (more than 90%) of the wild coconuts was more than 50% by weight, and the water content in the fruit was less than 15% of the total weight. Although the husk percentage could be approaching that of a more primitive form of niu kafa, the author was consistently designating them as niu kafa.

It is tempting to assume that these populations of semiwild, self-sown coconuts could have formed the initial material for selecting ennobled forms of coconuts having bigger nuts and higher kernel content. The Philippine coconuts are generally big in size, e.g., San Ramon, Philippines, Ordinary (Table 3.2).

Notwithstanding what has been stated earlier, generally larger-sized nuts contain more kernel and also more water. The present world germplasm collection being maintained in India does not contain any accessions selected for producing nuts with just high water content alone. In fact, in the literature, the author could not come across even any mention of forms having more water. Generally, bigger fruits contain bigger nuts, possess bigger cavities and, by inference, more water at tender nut stage (cf. data given in Rathnambal et al., 1995; also, Table 3.2).

2.2.4.3 India: Laccadive Islands

In India, the presence of naturally occurring wild coconuts has been reported from two locations—from some of the islands of Nicobar, both inhabited and uninhabited (8°0′N, 93°30′E) and Lakshadweep (10° N, 72°30′E). Even though this fact had been known from 1940AD, only few little studies have been done on them so far (Balakrishnan & Nair, 1979; Jacob & Krishnamoorthy, 1981; John & Satyabalan, 1995; Krishnakumar, Jerard, Josephrajkumar, & Thomas, 2014; Krishnamoorthy & Jacob, 1981; Samsudeen, Jacob, Niral, et al., 2006; Samsudeen, Jacob, Rajesh, Jerard, & Kumaran, 2006; etc.). Even the scope of such studies has been very restricted so far; all of them looking at coconuts in general terms as source material for coconut improvement material and at varietal differentiation.

2.2.4.4 India: The Andaman and Nicobar Islands

This archipelago is situated in the Bay of Bengal. It lies as an arched string of about 320 islands of variable size stretching from Myanmar in the north to Sumatra (Indonesia) in the south (6–8°N, 92–94°E). Its northernmost end is only about 190 km from Cape Negrais of Myanmar, and the southernmost point of the Great Nicobar Island is just 150 km from Banda Aceh of Sumatra (Indonesia). Only about 30 islands are occupied by humans. The Nicobar Islands (8°0′N, 93°30′E) are an archipelagic chain in the eastern Indian Ocean (eastern Bay of Bengal). United Nations Educational, Scientific and Cultural Organization (UNESCO) has declared the islands as one of the World's Network of Biosphere Reserves. It is 1840 km^2 in area and has a population of about 37,000 people. These islands are part of a great island chain formed by the impact of the collision of the Indo–Australian plate with Eurasia (Wikipedia, 06 September 2015). They receive c. 3000 mm rainfall annually.

Balakrishnan and Nair (1979) found wild coconuts in several islands of the Nicobar Islands including some that were not populated by humans. They observed both talls and dwarfs (Table 8.4). They found great range

Table 8.4

(A) Variability in the naturally occurring coconuts in Nicobar Islands

Attributes	Talls	Dwarfs
Inflorescence length	110–150 cm	95–150 cm
Fruit color	Green, yellow, red	Green, yellow, red
Fruit size	600–8220 cc	2000–6000 cc
Nut size	300–2175 cc	500–1500 cc
Kernel thickness	9–16 mm	9–12 mm
Copra weight	80–450 g	80–250 g
Glucose content	2.4–6.2%	3–3.6%
Oil content	65–75%	65–70%
Nut (fruit) yield/year	0–201 nos.	26–125 nos.

(B) Fruit and nut characters

	Length/breadth ratio	
	Fruit	Nut
No. 15	1.9: 1.0	1.1: 1.0
No. 16	1.3: 1.0	1.1: 1.0
No. 17	3.7: 1.0	1.5: 1.0
No. 22	2.7: 1.0	1.6: 1.0
No. 23	2.0: 1.0	1.7: 1.0
No. 25	3.5: 1.0	3.0: 1.0
No. 26	2.2: 1.0	1.6: 1.0
No. 27	2.5: 1.0	1.8: 1.0

(A) Adapted from Balakrishnan, N. P., & Nair, R. B. (1979). Wild populations of Areca and Cocos in Andaman & Nicobar Islands. *Indian Journal of Forestry, 2,* 350–363. (B) Calculated/adapted from Balakrishnan, N. P., & Nair, R. B. (1979). Wild populations of Areca and Cocos in Andaman & Nicobar Islands. *Indian Journal of Forestry, 2,* 350–363.

of variability for various characters. For instance, the longest nut (fruit) measured 36 cm and the smallest 10 cm. They observed also several palms bearing 700 micronuts per year. Unfortunately, their measurements are not given.

2.2.4.5 India: Laccadive Islands

The second location in India where the coconut occurs in natural stands is the Laccadive Islands in the Arabian Sea (10°N, 72°32′E). They consist of 30 coral islands covering 12 atolls, 3 reefs, and 5 submerged sand banks. Only 10 islands are inhabited. The present population is over 64,000 people. The islands receive about 1800 mm rainfall annually. The main vegetation consists of naturally occurring coconuts (Wikipedia, 2015). The islands are located 200–440 km southwest of the coast of India in the Arabian Sea. The land area is 32 km^2 and the lagoon area is 4200 km^2. These islands form the northernmost part of the Lakshadweep–Maldives–Chagos group of islands (Chagos, 06°00′S, 72°00′E), which form the tips of a vast undersea mountain range, the Chagos–Lakshadweep Ridge (Wikipedia, 2015).

These islands have been known to sailors from the earliest historical period. It is mentioned in the 1st century text, *Periplus of the Erythrean Sea* (Casson, 1989).

Several studies had been done in the past on the variability of coconuts in the islands beginning 1950 CE (John & Satyabalan, 1955; Krishnamoorthy & Jacob, 1982; Samsudeen, Jacob, Niral, et al., 2006; Samsudeen, Jacob, Rajesh, et al., 2006; etc.). However, they have all been limited to studying the fruit component characteristics of the four major morphotypes described earlier with a view to using them in coconut improvement. These landraces show great variability in most of the morphological and fruit characters.

Jerard et al. (2016) surveyed about half the area of the relatively isolated island of Minicoy having naturally occurring coconuts. They studied 2672 coconut palms. They too grouped them broadly into four morphotypes—Laccadive Giant Tall (3%), Laccadive Ordinary Tall (82%), Laccadive Micro Tall (14%), and Laccadive Mini–Micro Tall (1%). The nut size of Laccadive Mini–Micro Tall may be the smallest size recorded for an extant morphotype, 4–6 cm only.

The average nut yields of the Laccadives are the highest among all the states in India. There is considerable demand for the small nuts and ball copra of the Laccadives in the mainland of India. There, the ball copra is used widely for making confectionary and bakery products and, to a limited extent, in cooking (confectionary items, mainly), especially in northern India.

We can notice also that the fruit and nut sizes of Laccadives coconuts are comparatively much smaller than those of other countries/regions. This indicates that over the centuries some conscious preference/selection have been exercised in retaining and perpetuating palms having smaller fruit forms in the islands. This is analogous to the possible selection for forms producing large nut size in the Philippines, as we have noted earlier.

2.2.4.6 Other Regions

Early European explorers, mariners, and naturalists have commented extensively about the natural occurrence of coconuts in thousands of islands that dot the Pacific (c. 25,000 nos.) and Indian (c. 10,000 nos.) Oceans. Some noteworthy accounts have been of Charles Darwin in the mid-19th century and the detailed plant-geographical account of Sauer (1967) of his studies in the Seychelles Islands. Accounts of numerous travelers have attested to the widespread presence of naturally occurring coconuts in most of the islands.

2.2.5 Commentary and Inferences

It is apparent from the foregoing account that for a long time, some form of selection and propagation for ennobled forms of the coconut appear to have been taking place in at least some regions of South and Southeast Asia including Malesia. South and Southeast Asia have been known as centers of ancient civilizations for the last 3–5 millenia (cf. Glover & Bellwood, 2004; Higham, 2004; Possehl, 2002; Thapar, 2002; etc.).

This is the inference that we can draw from the data available on naturally occurring coconuts from New Guinea, the Philippines, and India (Nicobar Islands, Laccadives) (Table 8.9). In the Philippines, the preference appears to have been for large fruits/nuts. Such fruits possess increased quantities of kernel, coconut water, and also husk. The climate in the Malesian region is most suitable for the coconut. In the Laccadives, the selections have been for smaller-sized nuts (Tables 8.5 and 8.6). In the Andaman–Nicobar Islands, the selection might have been for higher yields. This is just a tentative observation based on the very limited available data (Table 8.4). Nevertheless, an overall objective in all the coconut-growing countries (except the Laccadives) might have been for more and bigger nuts per palm for realizing higher yields.

Table 8.5 Characteristics of the important characters of the four main morphotypes of Laccadives

Characters	Laccadive giant tall	Laccadive tall	Laccadive micro tall	Laccadive mini–micro tall	F	CD
Plant height (m)	13.2	15.1	12.6	9.5	48.11	1.02
Bunch number/year	11.7	11.2	13.1	18.3	26.07	1.93
Female flower no./bunch	24.8	35.5	117.7	105.0	49.21	20.60
Fruit number/bunch	6.9	11.6	69.5	9.9	227.40	6.07
Fruit weight (kg)	1.15	0.56	0.27	0.31	339.71	85.69
Husk thickness (cm)	4.7	4.2	3.0	0.8	1183.21	0.15
Shell thickness (mm)	3.5	3.8	3.0	1.5	541.91	0.13
Inner cavity volume (ml)	242.0	50.3	14.8	1.1	1053.12	10.50
Shell weight (g)	167.6	92.8	46.1	3.6	871.52	7.24

Jerard, B.A., Rajesh, M.K., Thomas, R.J., Niral,V., & Samsudeen, K. (2016). *The island ecosystems host rich diversity in coconut: Evidence from Minicoy Islands, India* (Submitted for publication)

Table 8.6 Characteristics of Laccadives coconuts

	Fruit length (cm)	Fruit breadth (cm)	Nut length (cm)	Nut breadth (cm)	Fruit weight (g)	Nut weight (g)	Kernel thickness (mm)	Shell thick (mm)
Jerard 2016 data								
LCGiant T	25.8	15.6	NA	NA	1149	558	1.3	3.5
LCT	20.7	20.6	NA	NA	564	261	1.4	3.8
LMT	16.3	8.1	NA	NA	269	122	1.3	3.0
LMMT	3.8	3.3	NA	NA	31	16	0.4	1.5
Niral 2016 data								
L Orange D	20.7	15.5	12.1	10.0	NA	NA	NA	NA
L Green Tall	20.8	16.2	10.9	9.4	NA	NA	NA	NA
Kaithathali	19.6	14.7	11.6	10.9	NA	NA	NA	NA
L Micro Tall	14.0	9.8	6.8	5.4	NA	NA	NA	NA
L Mini Micro Tall	4.6	3.5	2.7	2.0	NA	NA	NA	NA

We have even less data from Papua New Guinea. However, from the only descriptive account available of the region (Dwyer, 1938), it appears to have been similar to that in the Philippines and the Andamans (Golsen, 1989).

There has been as yet no evidence or reports from anywhere indicating that selections have been done for higher volumes of nut water.

With the information that we have presently, it is not possible to indicate any particular location in the world as the place of origin of the coconut. The present cultivated coconut is at best a semidomesticated species, or better stated, a semiwild species, as we shall see later. The initial steps for ennoblement might have been going on independently in all the regions of the world to varying degrees and with different objectives, where the coconut has been an important component in the lives of the people. But the initial primitive coconut appears to have looked broadly similar everywhere, having long fruits, high husk content, and smaller fruits and nuts, as had been suggested by Harries (1978). However, they were much more primitive than what he had designated as niu kafa. Hence, the coconut appears to have had a diffuse origin in this region both in time and place—in the Indian Ocean islands, insular Southeast Asia (including New Guinea) and western Melanesia including the Philippines—and even objectives. Several early authors (Beccari, 1917; De Condolle, 1886) had indicated one part or the other of this vast area as the center of domestication of the coconut. This is the inference that we can arrive at with our present understanding of the situation.

3. THE ANCESTRAL SPECIES

3.1 Inferences From Experimental Studies

As of 2016, no experimental work specifically designed to determine coconut origins has been done so far. At the same time, the phylogenetic studies done on the family Aracaceae, tribe Cocoseae, and subtribe Attaleinae, have thrown some light on the origin and relationships of the genus *Cocos*. The work on this aspect has been reviewed in Chapter 5. The results are summarized in Table 8.7.

The genus *Cocos*, to which the coconut is now assigned, has been a monotypic species since the last 100 years now, after Beccari (1916) transferred all the species except the coconut, *Cocos nucifera*, to the other genera, *Butia* and *Syagrus*. Prior to that, the genus *Cocos* used to be "a dumping ground" (Glassman, 1987) of species/taxa from other genera. No genetic studies involving other genera have been carried out in the genus *Cocos* to understand its relationships with the other genera. The few phylogenetic studies that have been done have given

Table 8.7 Comparison of six subtribe Attaleinae genera including *Cocos**

	Cocos	Attalaea	Jubaeopsis	Parajubaea	Syagrus
Species no. & chromosome no.	One sp. 2n = 32	c. 74 spp. 2n = 32	One sp., 2n = 160–200	3 spp. 2n = not known	31 spp. 2n = 32
Genome size	5.996 pg	4.02–4.34 pg	2.98 pg	Not studied	3.9–6.2 pg
Distribution	Worldwide, tropical, subtropical	Mexico, Caribbean, Bolivia, Peru	Southeastern S. Africa, rare	Ecuador, Colombia, Bolivia at high altitudes	Caribbean, S. America, mostly Brazil, in drier parts
Ecology	Strand plant; in humid weather, up to 900 m	Wide habitat range: Tropical rain forests to dry savannahs	Coastal reaches, gregarious, rocky banks of rivers	Humid ravines of cooler sandstone mountains at high altitudes, 2400–3400 m	Extremely variable, dry to semidry areas, acaulescent, conspicuous in dry vegetation types
Habit	Solitary, moderate, unarmed, pleonanthic	Solitary; small to massive	Clustering, branching at base with short erect stems, pleonanthic	Solitary, tall, stout or rather slender; unarmed, pleonanthic	Solitary or clustered, pleonanthic, small to tall stem
Fruit size, shape	Very large, 20–40 cm long; ellipsoidal to broadly ovoid, one seeded	9–10 cm long, 1-several seeded	c. 3.3 cm long; one seeded; globose with apical beak	5–6 cm long, oblong–ovoid, beaked; 1–3 seeded	2–5 cm long, small to relatively large; one, rarely two-seeded
Mesocarp	Very thick, fibrous, dry	Fleshy or fibrous	Thick, fibrous; slightly fleshy	Thin, fibrous	Fleshy or dry edible
Endocarp	Thick, woody	Very thick; strong, smooth	Thick, bony, with 3 vertical grooves	Thick, very hard, with 3 prominent ridges	Thick, woody

Endosperm	Homogeneous; large central cavity with partly filled water	Solid	Homogeneous with large central cavity, may be eaten	Homogeneous, hollow	Homogeneous, ruminate central cavity in some species
Leaf	Pinnate, neatly abscising	Pinnate	Pinnate in 5 vertical rows, marcescence or nearly abscising	Pinnate	Reduplicately pinnate
Stamens number	6	3–75	7–16	c. 15	6
Inflorescence	Monoecious; solitary; interfoliar; auxiliary; branched to 1-order	Monoecious; staminate or pistillate or both on same inflorescence	Monoecious; solitary; interfoliar; branching to 1-order, protandrous	Monoecious; interfoliar; erect, or pendulous; 1–2 order branching	Monoecious; solitary; interfoliar; rarely spicate, usually branching to 1-order, protandrous (?)
Relationship	Moderately sister to *Parajubaea*, *Attalea*, and *Syagrus*	Sister to a clade of *Lytocaryum*; subclade of *Syagrus* with moderate support; monophyletic with high support	Sister to subtribe Attaleinae except *Beccariophoenix*	Moderately sister to *Cocos*	Polyphyletic

With *Lytocaryum* included (Noblick & Meerow, 2015).
Various, chiefly Glassman, S. (1987). *Revision of the palm genus Syagrus Mart, and other selected genera of the Cocos Alliance* (230 pp.). Urbana, IL, USA: University of Illinois Press, Noblick, L. (2013). Syagrus. An overview. *The Palm Journal, 205,* 4–30, Noblick, L. R.., & Meerow, A.W. (2015). The transfer of the genus Lytocaryum to Syagriss. *The Palms, 59,* 57–62. Dransfield, J., Uhl, N.W., Asmussen, C. B., Baker, W.J., Harley, M. M., & Lewis, C. E. (2008). *Genera Palmarum. The evolution and classification of palms* (732 pp.). Kew, UK: Kew Publishing, and Gunn, B. F., Baudouin, L., Buele, T., Ilbert, P., Duperray, C., Crisp, M., et al. (2015). Ploidy and domestication are associated with genome size variation in palms. *American Journal of Botany, 102,* 1625–1633.

different and inconclusive results, especially at the lower taxonomic levels. The three genera that have shown sister relationships with the genus *Cocos* have been *Parajubaea* (Baker et al., 2009; Gunn, 2004), *Syagrus* (Meerow et al., 2009), and *Attalea* (Baker et al., 2011; Meerow et al., 2014). In addition, Beccari (1917) had proposed that *Jubaeopsis caffra* has a close relationship with the coconut.

A comparative look at 12 characters of the genus *Cocos* with those of the four genera—*Attalea*, *Jubaeopsis*, *Parajubaea*, and *Syagrus*—does not give much indication about the genus that may have the closest relationship with the coconut. Two of the genera possess the same chromosome number as *Cocos*, $2n = 32$. The chromosome number of a third one, *Parajubaea*, is unknown, and that of the fourth genus, *Jubaeopsis*, is variable, $2n = 160–200$ (Table 8.7). One genus that is as monotypic as *Cocos*, *Jubaeopsis*, is endemic to a river valley in South Africa. The other genera are Neotropic in distribution. Although *Attalea* has c. 74 species and *Syagrus* 31 species, *Parajubaea* has only three species. *Syagrus* has the same chromosome number as that of *Cocos*; in the other three genera, it is variable, ranging from 3 to 75. Some species of *Attalea* and the lone species of *Jubaeopsis* show ecological similarity with *Cocos*; most species of other genera seem to occur largely in dry habitats. At the same time, the two genera Attalea and Syagrus share the same chromosome number with the coconut ($2n = 32$). Further, some of their species too show some morphological and ecological affinities with the coconut.

With the very wide variability shown especially by *Attalea* and *Syagrus*, we may have to compare the monotypic *Cocos* with individual species of both these genera to seek leads about their relationships.

In the phylogenetic studies, as we had noted earlier, the results about the relationships have varied in different studies, especially at the lower taxonomic levels, with changes in the species (both in the number and the species used) chosen from different genera, the number of species and coconut varieties used, and also the method of analyses used by different groups. The conclusions that the different authors have arrived at are based both on their own studies and those of pooled analyses. Such variations in results may be attributable to variations in the Materials and Methods used by different authors. In the case of the coconut too, the number of samples used in different studies have ranged from only one to six in various studies. In the two genera, *Attalea* (c. 74 species) and *Syagrus* (31 species), the number of species selected have varied substantially in different studies. An instance in point is that Meerow et al. (2009) had found a sister relationship between *Syagrus* and *Cocos*, and in Meerow et al. (2014), they recovered sister relationships between *Attalea* and *Cocos*. This has been attributed to incomplete lineage sorting. Another problem is that generic delimitation appears to have been done rather broadly in at least the tribe Cocoseae. Much interspecific variability is apparent within the genera that contain relatively large numbers of species such as *Syagrus and Attalea*. Analogous results have been obtained by Baker and associates in their 2009 and 2011 studies. But, in their second study, Baker et al. (2011) had additionally pooled the data of all the earlier phylogenetic studies done thus far in the subfamily Arecoideae by different authors and carried out a combined analysis of the entire data set. This made it possible for every genus in the subfamily to be represented by at least one DNA sequence. This could have made the study, and hence the results, more robust.

3.2 Results of Phylogenetic Studies

With regard to suprageneric differentiation and dispersal (Baker & Couvreur, 2013a, 2013b), the earliest divergence (crown node) of palms occurred in Laurasia (Eurasia plus North America). Then, palms dispersed on a number of occasions into all major tropics. A dispersal westward into the Pacific Ocean, east of the Wallace Line (Fig. 7.1), took place on multiple occasions beginning 50 Mya. The authors proposed a complex pattern of westward expansion into the Pacific and also extinction in the region. Many of the migrations that appeared to have happened predate the Miocene (23–5 Mya), whereas within the tribe Cocoseae, the stem-node divergence of subtribe Attaleinae (56 Mya) also occurred in South America. The subtribe then expanded into the Indian Ocean islands 36 Mya, and into Africa, after 29 Mya. Within the subtribe Attaleinae, *Cocos* was positioned as sister to *Attalea*. The stem node of *Cocos* and *Attalea* was optimized ancestrally in the Atlantic coastal forest region. The authors had also indicated their inability to account for the entire paleodistribution of the tribe Cocoseae members.

Baker and Couvreur have not explained the possible manner of long-distance dispersal of the tropical groups. One possibility was dispersal via the Antarctic. This might have happened during 55–40 Mya. In this context, Lear, Bailey, Pearson, Coxall, and Rosenthal (2008) had indicated that the Eocene temperature in Antarctica was cool and temperate, and hence, amenable to the growth of higher plants. The dispersals into Eurasia, the Indian Ocean, and the Pacific (7–8 events each per region) occurred at short intervals during Early Eocene (56–47.8 Mya) to Late Miocene (7.2–5.3 Mya).

Interestingly, Meerow et al. (2014) too had obtained broadly similar results. They too had found that the crown node of the American Attaleinae was optimized ambiguously in the Atlantic coastal forest region. From the crown node of *Cocos*, as many as 14 dispersals possibly took place. The genus *Cocos* was placed as sister to a *Lytocaryum/Syagrus* subclade (Incidentally,

Noblick and Meerow (2015) have suggested the merging of *Lytocaryum* with *Syagrus*). The two-tree estimations of Meerow et al. and the Bayesian analysis of the supermatrix had positioned *Cocos* as sister to *Attalea*. They had found that just four additional base substitutions were enough to support this resolution. They pointed out the similarities between the two genera in that both of them possessed fibrous mesocarp.

Incidentally, both Baker and Couvreur (2013b) and Meerow et al. (2014) have observed a wide time lapse of 20–25 Mya between the crown and stem in the divergence of subtribe Attaleinae. Although Baker and Couvreur did not comment on this aspect, Meerow et al. (2014) attributed this event to the extinction of nearly one-third of the terrestrial vegetation and consequent great decline of species abundance during Late Tertiary because of the impact of an asteroid on the Earth (Chapter 6). One point supportive of this proposition—though the authors have not stated it so—is the report of Gomez-Navarro, Jaramilo, Herrera, Wing, and Callejas (2009) from northeast Colombia ascribed to the Middle–Late Paleocene.

There is also a report of the finding of an extinct Attaleinae species by Dransfield, Flenley, King, Harkness, and Rapu (1984) from Easter Island in the eastern Pacific (27°00′S, 109°00′W), and its age is only 900 CE.

It may also be worth considering how we could extrapolate the reports of the recovery of *Cocos*-like stem (Sahni, 1946), and the mostly petrified fruits, by various authors from the Deccan intertrappean beds of central India by various authors (Table 8.1). The fruit size of at least three of them come within the range of variability of the present-day coconut fruit. Please refer also to the finding in Rigby (1995) from the Chinchilla Sand Formation (Queensland, Australia). This too shows similarities in size—and with some changes—age also. They point out a more primeval form than the primitive forms of coconut observed by Greuzo (1990) in Eastern Samar (The Philippines) and Dwyer (1938) in Matty Island, Papua New Guinea. If we take into consideration together, the presence of very small fruits—smaller than the even more primitive form of Dwyer (1938); some of the material described from western central India by various authors; from Queensland, Australia (Rigby, 1995); and the bearing coconuts from the Nicobar Islands and Laccadives (Balakrishnan & Nair, 1991)—the jigsaw puzzle about the immediate primitive taxon of the coconut could possibly be put together. We will need to compare them with the species (and genera) described from subtribe Attaleinae, that have been reported to show affinity with the genus *Cocos* by various authors—*Parajubaea*, *Jubeopsis*, *Syagrus*, and *Attalea*.

It has been well documented that the family Palmae is characterized by high levels of endemism and variable growth habits and habitat preferences (Dransfield et al., 2008). Perusing the keys to the species of these genera as well as the species descriptions of some of the genera (cf. Glassman, 1987, 1999; and earlier), etc., will show that the various species within even the same genus show more morphological "discreteness"/differences and fewer morphological similarities among them than usually seen in the comparative morphologies of related species and genera of other plant families.

3.3 Inferences

In these circumstances, it is difficult to zero in on any species or, for that matter, even genus that could be identified as the progenitor taxon, or one having a sister relationship with the coconut. For the present, we may assume it to be *Attalea or Syagras* based on the results of both Baker et al. (2011) and Meerow et al. (2014), and possibly also *Jubaeopsis* or *Parajubaea*.

All that we can indicate at the present time is that the naturally occurring primitive forms of the coconut that occur in the midwestern Pacific region, comprising western Melanesia, eastern Philippines to Papua New Guinea—and also the islands and archipelagoes of the Indian Ocean—are the immediate ancestral material of the present-day coconut, a form that is much more primitive than niu kafa as defined by Harries (1978). Nuts similar to these, but much smaller in size, have been recovered from the Society Islands (Lepofsky et al., 1992). They "neither matched the extreme wild or domesticated form of coconut having thick husk, oblong shape, and moderately thick-shell." The authors carbon-dated the two nuts to 1270 ± 60 BP and 1360 ± 70 BP. Because one of the authors (HH) is very familiar with the coconuts, we may accept their conclusions at face value. The fossil nuts (fruits) recovered from Queensland (Australia), and western central India, and the primitive coconut palm populations observed in the Philippines, Papua New Guinea, Nicobar Islands, and Laccadives also appear to come within this category.

It is tempting to suggest that the nut material that Spriggs (1984) obtained from Mo'orea, Vanuatu, is similar to what has been obtained by various authors from both North Island and South Island, New Zealand. They varied in length from 3.5 to 5.00 cm. Ballance, Gregory, and Gibson (1981) had obtained their samples from proximal turbidites (a sediment or rock deposited by a turbidity current). At the same time, Endt and

Hayward (1997) proposed that their material showed the appearance of *Parajubaea*. The suggested age was Late Tertiary (66.0–23.0 Mya); Incidentally, the nomenclature of the Tertiary has been now changed [It is now divided into two periods, Neogene (23.0–2.6 Mya) and Paleogene (66.0–23.0 Mya)] (cf. p. x) The samples obtained from New Zealand were often in good numbers and from scattered locations (1 km or more apart). Could they all have come from tsunamis, or could they be taken to indicate that some genera of the present subtribe Attaleinae—say, even *Parajubaea*—was present in the South Pacific in the former Tertiary period? There is some comparability in the time of differentiation of subtribe Attaleinae and the time and context suggested for the New Zealand *Cocos*-like fossils (Tables 8.1 and 8.7). The biogeographical studies of Baker and Couvreur (2013a, 2013b) indicate that the subtribe Attaleinae, to which the genus *Cocos* belongs, expanded into the Indian Ocean islands around 36 Mya and into Africa after 29 Mya.

With one or two exceptions, the fossil fruits reported from the Deccan trap regions of central India come broadly within the range 9–11-cm length. This matches in size with the fruits of *Attalaea* of the *Cocos* Alliance/subtribe Attaleinae. The genus *Attala* of this subtribe has been found by Baker et al. (2009) and Meerow et al. (2014) as having a sister relationship with *Cocos*. Even the size of the primitive forms (Mini–Micro) described from the Laccadives and Nicobar Islands are smaller than this (9.5–10.5 cm).

4. TIME OF ORIGIN

4.1 Historical

The time of origin of a species is indicated best by archeological and paleo-botanical evidences. We have reviewed these aspects already vis-à-vis the coconut (Chapter 4). Tables 8.1 and 8.2 give summary results. We have observed before about the inherent limitations and difficulties of studying temporal aspects of origin. Hence, the findings given here need to be viewed with some caution.

The earliest archeological record to date of coconut seedling-leaf samples has been recorded from Pagan, Mariana Islands (15°0′N, 145°0′E) (Fosberg & Corwin, 1958) which has been dated to Late Quaternary (1.6 Mya). All the other archeological material belonged to the Holocene epoch (from 10–12,000 Ya to the present). The most recent one has been from Moorea, Society Islands, Polynesia (Lepofsky et al., 1992; Table 8.2).

Their dates were c. 1210–1410 BP. They consisted of entire coconuts (Fig.). The earliest archeological remains of coconut fruits are those obtained from Ancilum Island, Vanuatu, dated to 5420 ± 90 BP [Spriggs, 1984 (Table 8.2; Fig. 4.8)].

There are, however, only a few (five) unequivocal archeological records of coconut remains from the Pacific region, and none from any other region, including South and Southeast Asia. In this context, Kirch (1997) has observed that "coconut shells preserved in swamps and coconut pollen identified from pollen cores indicate that the plant was *naturally distributed* (italics by author) as far east as the Cook Islands (and probably also the Society Islands) prior to human dispersal in the Pacific … Most earth-oven features contain traces of carbonized coconut endocarp. The very hot, intense ignition of these dried shells is still the preferred method of lighting earth ovens throughout most of tropical Oceania."

Recapitulating (cf. Bellwood, 2013): present-day humans, *H. sapiens*, migrated out of Africa into Eurasia 135,000 to 90,000 Ya, and the modern hunter–gatherers had reached the Earth's habitable limits by 45,000 Ya. They reached South Asia 75,000 Ya and the tropical Sundaland also by this time along with middle Paleolithic technology. During the Last Glacial Maximum (LGM: 25,000–19,000 Ya), Sundaland was c. 2.2 M km^2 in area (Fig. 7.1). After the LGM, when warming began, sea level rose up to 120 m (c. 90 m according to Clark et al. (2012) during 15,000–7000 Ya) to reach the present level. This sea level rise was evidenced most in Southeast Asia, but the effects were apparent elsewhere as well. For instance, as recently as c. 8000 Ya, Australia and New Guinea constituted one landmass, and Sri Lanka was a part of the Indian subcontinent. The present-day humans reached Australia and New Guinea entirely through the tropical world across the Arafura Sea.

New Guinea and Australia had remained as a single land mass across the Torres Strait (between Papua New Guinea and Cape York Peninsula, Queensland, Australia) until c. 8000 Ya. There are evidences of human presence in Sahul 20,000 Ya, but only up to Bismarck Archipelago, Solomon Islands, and Admiralty Islands (and not beyond; i.e., from north of Papua New Guinea). Evidence of human occupation in several caves 40,000 Ya has been obtained in the Mollucas, Sulawesi, Timor, and Taland. There were no humans beyond this limit in the Pacific Ocean. Only in more recent times, the ethnolinguistic group, Austranesians, had traveled 4000 Ya, across the sea gap of c. 1000 km from Taiwan to Batan Islands (20°30′N, 121°50′E, north of the Philippines). It is unclear, if they went intentionally, or that it was a landfall (Bellwood, 2013).

4.2 Food Production System in West Pacific Ocean

Archeological evidences have shown that west Pacific Ocean had its own unique food production system. Although the food of the people consisted primarily of marine and animal resources, plant foods too were used. Their collection and production were based on horticulture and arboriculture. Some specialists have proposed that vegetatively propagated plants such as sugarcane and bananas, and possibly also taro, originated in New Guinea. Their origins are dated to 4000 BCE (Allen & Gosden, 1991; Bellwood, 1987, 2004; Kirch, 2000; and others). The other crops that were important components of this production system were breadfruit (*Artocarpus altilis*), coconut, sago palm (*Metroxylon* and several others), and *Canarinm* spp. (Pacific chestnut), etc. This food production system had evolved independently from the east Asian production systems which had been based on Asian rice and millets. In New Guinea highlands, an advanced system of irrigated crop culture, consisting mainly of banana and taro, initially developed about 6500 Ya. However, there were no domesticated animals. This development led to the colonization of the Pacific Ocean islands, as we will see subsequently.

4.3 Colonization of Oceania

Traditionally, Oceania is divided into Micronesia, Melanesia, and Polynesia. In recent times, to avoid racial connotation, and to better align itself with the language families distribution, a new classification, Near Oceania and Far Oceania, is being increasingly used (Fig. 7.4).

A couple of millennia after the food production system based primarily on rice and millets developed in northeast China, a human migration began to take place toward the south. From south China, one group moved to Taiwan c. 5500 Mya, and another group toward the south to Vietnam, and further onto the Malaysian Archipelago. They had carried with them basic skills in farming and cooking, which had been based on rice and millets. This ethnolinguistic group has since been known as Austranesians. They had in the meantime developed skills in seafaring. What is relevant to us is that a part of the group that settled in Taiwan crossed the sea c. 4400 Ya, to the northern Philippines and, all the while, undergoing steady transformations in their lifestyles (Bellwood, 2013). From the Philippines, one group moved across the sea to western Micronesia (Fig. 7.5).

The Austranesians in eastern Indonesia, east of Sulawesi, underwent another major transformation after c. 3500 BCE. They replaced rice and millets with tuber and tree crops, including breadfruit and coconuts. After this major change, migrations of Austranesians took place into Oceania after 2300 BP. They had also developed a distinctive style of pottery. And, together with the arboriculture, tuber crop cultivation production systems developed into a new culture known as Lapita culture (Allen & Gosden, 1991; Kirch, 1997; etc.).

The Austranesians had developed skills in sea voyaging, as evidenced earlier from their travel from Taiwan to north Philippines (c. 1200 km), and later, further to Mariana Islands in Micronesia (c. 2300 km). In addition, they possessed maritime subsistence skills. Further, they domesticated animals such as pigs, dogs, and fowl. Beyond Southeast Asia, the Austranesians, including the Lapita people, were inveterate sailors. They used to move in boats consisting of dug-out canoes, double sailing canoes, double outriggers, and tied log piles using sennit (braided cordage in flat or round form using a variety of fibers including various hemps, and also coconut fiber). Sennit is not a synonym of coir, as some authors have made out.

In Micronesia, the first settlers had moved into the Palau Islands (Micronesia) in 2000 BP, and the Carolinas, consisting mostly of only atolls, by 1500 BP. They reached there directly from the Philippines across 2300 km of open sea (Bellwood, 2013; Kirch, 2000). Between 3350 and 2900 BP, these Neolithic colonists occupied Melanesia (excluding New Guinea) and west Polynesia. Beyond (and including) New Caledonia, the Lapita culture people were the first modern humans, *H. sapiens*, to occupy Vanuatu, Fiji, Tonga, and Samoa. There were no humans in near Polynesia before the arrival of these Neolithic Austranesians c. 3000 Ya.

The settlement of central and eastern Polynesia—Marquesas, Societies, Cooks, Australs, Tuamotu, Hawaii, even Easter and New Zealand—occurred after 700–1000 CE. The exact reason for this long break or pause is unclear, but Bellwood (2013) has suggested that many oceanic atolls could have been possibly below the elevated sea levels until about 1000 CE.

4.4 Time of Origin of Coconuts

We have seen before that the few drill cores and archeological sites in the Pacific Ocean islands had shown the wide presence of *Cocos*-like pollen and remnants of whole coconuts from a few sites (Table 8.2). In addition, Kirch (1997, 2000) had observed that coconut shell bits and burnt coconut remains used to be commonly recovered from numerous archeological sites of Polynesia. All of them indicate the wide presence of naturally occurring coconuts in the Pacific Ocean islands, more in the west and central Pacific long before the arrival of modern day humans, *H. sapiens*. In addition, even though direct evidences are not available, all the Pacific Ocean specialists have observed that the Austranesians (including the Lapita people) must have been carrying coconuts with them, mainly for drinking water during their voyages, because many of the islands and atolls, and particularly the smaller ones, do not usually possess native sources of water even for drinking. Added to these evidences are the fossil reports, especially from New Zealand and South Asia (India). We know that coconuts formed an integral component of the Pacific food production systems (Yen & Mummery, 1990; and others), but of lower importance, even among vegetable sources. After the marine and land animal resources, which used to constitute the main source of food for the Oceanians, the major plant sources used by them were breadfruit, pandanus, sweet potato, and sago palm. Some others like the coconuts, *Barringtonia* spp., Polynesian chestnut (*Inocarpus*), etc. too were being used (Allen & Gosden, 1991). These indicators lead us to propose that the Oceanians must have been practicing at least a modest-level ennoblement of primitive coconuts, *Cocos nucifera*, even beyond the western Pacific and New Guinea. The available evidence indicates that this process of ennoblement of the primitive coconut, *Cocos nucifera*, must have been a diffuse process in space, time, and objectives; objectives, because the coconut is a multipurpose tree.

A couple of authors have estimated the time of differentiations of the genus *Cocos*. We have covered this in Chapter 5. The information has been given also in Table 8.7. The estimates for the date of differentiation of the genus *Cocos* are 26.71–22.20 Mya (Gunn, 2004), 44.4–23.9 Mya (Meerow et al., 2009, 2014), and 23.61–3.92 Mya (Baker & Couvreur, 2013a, 2013b). All of these estimates cover long time spans, and we do not know how the primitive coconuts would have looked like then and what their characters would have been. If we go by the images and descriptions of some authors (Rigby, 1995), several (one from the Deccan traps) of the fossils of the *Cocos*-like material showed considerable variability in at least the fruit characters. We cannot help but observe that the taxonomical delimitations within the tribe Cocoseae/subtribe Attaleinae appear to have been in a nebulous state.

We need also to reckon that the coconut is a multipurpose tree. Every part of the coconut palm can be put to active economic use. This was so in fact in many regions until about 60Ya. Nevertheless, the emphasis in the part used in the coconut appears to have been different in different regions of the world.

Generally, in most crop plants, there used to be one primary objective in the process of domestication. The classical instances of this has been the resistance to shedding of grains in the cereals. In perennial crops, the best examples of primary objectives are the selection for seedlessness in bananas and breadfruit (*Artocarpus atilis*). Early humans did not appear to have had any such primary objective for the coconut. For one, the coconut was just one among a bouquet of plants available to the early humans as a source of vegetable food. Besides, it was always considered as a multifunctional plant. However, in Polynesia, the coconut could have been crucially important as a source of water. For we now know a bigger fruit and nut size confers higher volumes of water, and also more kernel and husk (Tables 3.2 and 8.9). But elsewhere, the primary objective might have been different in different regions. It might have been for bigger and bolder nuts/fruits, higher fruit numbers (overall "nut" yields), higher kernel content, and so on. Selection for more water content alone is a moot point, because at the tender nut stage (3–5 months age), the nut cavity is filled with water irrespective of the nut size. Hence, selection for bigger fruit size would have achieved the twin purposes of high water and husk content at the tender nut stage and higher kernel content and more water at the mature nut stage as well.

In this context, it may be worthwhile quoting Nayar's (1978) observations: "there is no need to assume the existence of a truly wild coconut palm either now, or in the immediate past, since as with many other tropical plants used by man, the present-day coconut palm does not also appear to have made any significant evolutionary advancement as a result of deliberate cultivation by man … The germplasm collections made from around the world, including the numerous islands of the South Pacific and Indian Oceans, do not show the presence or prevalence of genuinely primitive and advanced characters for even the economic attributes."

5. MODE OF EVOLUTION OF THE COCONUT

5.1 General

This is the fourth and final aspect relating to the origin of the coconut. Hardly any author has commented on this aspect so far. No experimental work has yet been done on this aspect.

We need to keep in mind a couple of points when we consider this topic. The first one is that the coconut, *Cocos nucifera*, has been a monotypic species since 1916. Since then, both classical taxonomists and molecular phylogeneticists have endorsed this status. Second, the various phylogenetic studies have indicated that the genus *Cocos* differentiated during 30–22 Mya (Table 8.7). Third, we do not yet know for sure about the nearest related species of the coconut, *Cocos nucifera*. We can, of course, safely assume that the immediate progenitor of the present-day cultivated coconut—which we have termed here as semiwild in form—must have been akin to or the same as the "micro" forms reported from the various Indian Ocean archipelagoes and also New Guinea.

With our present understanding of the subject, the most closely related species might have belonged to one of four genera, *Attalea* (c. 74 species), *Parajubaea* (3 species), and *Syagrus* (31 species), and, *Jubaeopsis* (1 species)—but, more probably, *Syagrus* or *Attalea*. However, among these, the two most recent phylogenetic studies have indicated the closest sister relationship with *Attalea* (Baker et al., 2011; Meerow et al., 2014) followed by *Syagrus* and *Parajubaea*, in that order. Whether it is extant or extinct, we do not know about the most closely related species.

5.2 The Domestication Process

Evolutionary biologists, archeologists, and anthropologists agree that the domestication of all the plant and animal species was carried out only by present-day humans, *H. sapiens*, and as recently as during the Holocene period, i.e., only within the last c. 12,000 years. Prior to that, they (*H. sapiens*) were hunter–gatherers. They began the process of domestication of animals and plants, only after they adopted an at least seasonal sedentary way of life. Extensive documentation is available on this process. Hence, we shall not go further into this aspect.

Another basic aspect of domestication is that the early humans took to this process of cultivations/farming, when in a location, they had found that the quantity of plants and animals available for foraging in nature (in the wild) was insufficient to meet the community's normal needs. In most areas of the Pacific and Indian Oceans, plant foods constitute only a small component of the food of the islanders. In the coconut, such a situation as the need for domestication did not appear to have occurred in at least the Pacific islands. Even presently, this continues to be so in several countries/regions as the Maldives, Laccadives, Nicobar Islands, the Pacific Ocean countries, etc. The local population is by and large able to meet all their needs from the naturally occurring coconuts around them to undertake two tasks—to sow/plant seeds of the wild forms in areas that they felt were suitable for propagation and, secondly, to select forms giving higher yields and/or possessing superior attributes. This may also explain the reason for the coconut continuing to remain in a semidomesticated/semiwild state. Some regions—such as the Kerala–Konkan coast of India, Sri Lanka, the Philippines,

Southeast Asia—where the population was high, the use of the coconut was intense not only for the kernel, but also for its other plant and fruit components, and the supplies of naturally occurring coconuts might have been insufficient to meet their full daily needs, consequently, they might have been compelled to take up the planting of seed nuts collected from superior–looking palms, and also giving them some protection. This practice continues even today in many regions of South and Southeast Asia. This process can be seen in the Maldives and Andaman and Nicobar Islands today.

Worldwide, the coconut scenario has undergone a major transformation beginning about the mid-19th century. Then, the global demand for coconut oil began to increase substantially both for industrial and food purposes. This set the stage for the golden age of the coconut. This continued till the mid-20th century. During this period, coconut oil had become the most traded vegetable oil in the world, as we had noted earlier. Consequently, this led all the European colonial countries having overseas colonies in the tropical regions to set up commercial coconut plantations. This process incidentally must have led to at least some extent of ennoblement, as all the European colonial countries were setting up plantations using only selected nuts from selected superior palms. This golden age ended by the 1950s, as other vegetable oils began to displace coconut oil. By 1961, when the FAO began publishing production statistics of cultivated plants, coconut oil was already ranked only #3 in the world. In 2011, its ranking was a dismal #9. We have highlighted this in Chapter 1.

The earliest center of crop plants domestication was western Asia. This was based on wheat, barley, oats, etc. This was followed a couple of millennia later, in middle-eastern China based around the lower reaches of the Huang He River. This was based on rice (japonica) and millets. And then, a further millennium later, in the Indus Valley. This was centered around rice (indica), millet, pulses, etc. The other major and ancient centers of plant domestication have been in New Guinea and the Andean mountain regions in South America. In most of these centers (except in New Guinea and the Andean region, to some extent), the earliest domesticated plants were the cereals, including millet, followed by legumes and tuber crops. Domestication and/ or ennoblement of other crop plants were carried out much later.

In all the above crops, especially the cereals, millet, and seed legumes, the first and the most decisive domestication objective has been for a single character, viz, reduction in seed shedding. In wild plants, usually the maturing seeds are shed easily, and in a staggered manner. This is done to ensure better propagation and survival in nature. In most of them, this character (reduction in seed shedding) is governed by a single gene. Archeological evidence is available in several cereals and legumes to show that this indeed has been so—in diploid wheat, barley, oats, rice, etc. (Ladizinsky, 1999; Nayar, 2014a, 2014b; Zeven & Zhukovsky, 1975; Zohary, Hopf, & Weiss, 2012; and so on). With such a primary gain, the early farmers could seek further ennoblement for other desired characters such as higher yields, synchronized periods of maturity, early maturity, improved cooking quality, selection for biotic and abiotic characters, and so on, and approximately in that order. In certain vegetatively propagated plants too, the initial selection objective in domestication appears to have been for a single decisive character. This may be seen most obviously in banana (seedlessness; Nayar, 2010; Simmonds, 1966), breadfruit (*Artocarpus subtilis*) (seedlessness; cf. Rajendran, 1991), reduction in antinutritional factors (low cyanic acid content in cassava, reduced acridity in aroids, etc.), and so on.

5.3 Domestication Process in the Coconut

In the coconut, however, the situation is different, and in this respect, unique. This is apparently because the coconut is a multipurpose crop plant. Every part of this palm has been used traditionally and actively for different purposes—like roots for fuel and folk medicine; trunk for house building; canoe and boat building (in the Indian Ocean islands); leaves for thatching and fuel; even sails and floor matting; the nuts for water; coconut kernel for food; and oil for cooking; as emollients in folk medicine; the shell as container and as excellent fuel and for industrial purposes; the husk for extracting fiber (which is used for making coir, canoe building in sewn boats as used in the Indian Ocean); tender leaves and inflorescence for religious and social functions, and so on. A major change in this use pattern began to take place after the Second World War, and a second equally major transformation in habits, but of a more comprehensive nature, since the last three decades. Readers are invited to refer to Chapter 2 for further details and references.

   In the coconut, early modern humans (*H. sapiens*) appear to have been using coconuts for two main purposes, but in different regions. With the very limited data now available, coconut water was being used for drinking, especially during their sailings in the Pacific and Indian Oceans—significantly more in the Pacific Ocean islands—and coconut kernel as food. Although for water, they would have had to carry tender nuts of 4–5 months' age for use as food and oil, they would have had to use mature nuts of 10–12 months age. Only the mature nuts are able to germinate and produce new seedlings. Hence, the first farmers and settlers—first the Austranesians, including the Lapita people—especially in the

Pacific islands, might have had ambiguous or diffuse objectives for the ennoblement of the coconut. Added to their burden or difficulties in setting a clear objective for ennoblement is the fact that selection for these factors would have required them to perform a laborious and time-consuming method to identify palms possessing these superior characteristics, and that too after developing a settled life.

The coconut is a perennial species and may need 20–25 years from planting to evaluate its performance: a tedious and time-taking process. At the same time, and fortuitously, the naturally occurring coconut—possibly because it is a predominantly cross-fertilized plant—shows much variability for all the potentially useful characters (Table 8.9). Still, it may not be erroneous to assume that the early communities including those who were active seafarers would have attempted to carry out at least a low-intensity selection for some attributes. It is unlikely to have been for higher water content, for we know that all the tender nuts irrespective of their size and the number of fruits in a bunch are known to have their inner cavities filled with water.

Some supporting evidence for this proposition is available. One is the statement of Gunn, Baudouin, and Olsen (2011) that coconut did not appear to have undergone any significant level of ennoblement from the niu kafa traits. Equally significant is the observation of Lepofsky et al. (1992) that the nuts that they had recovered from the Society Islands (as late as 1270 ± 60 BP) were intermediate between the extreme wild (possibly the authors meant the niu kafa type) and domesticated forms. This statement corroborates Gunn et al.'s (2011) suggestion that only partial ennoblement had taken place in present-day coconuts. This is what we have been stressing all along in these pages.

5.4 Selection Under Natural Conditions

The term "selection under natural condition" was coined independently by Charles Darwin and A.F. Wallace. "Selection" is the process determining the relative share allotted to individuals of genotypes in the propagation of a population. Natural selection is the differential fecundity in nature between members of a species possessing adaptive characters and those without such advantages (King, Stansfield, & Mulligan, 2006). Natural selection alters gene frequency in wild populations through differential fertility and survival of various genotypes. In most cases, natural selection is expressed by differential contribution of various genotypes to the next generation (Ladizinsky, 1999).

Evolution by natural selection is a process occurring in nature. It is meant to ensure the survival of the fittest in nature under changing natural conditions. Such forms are unlikely to possess characteristics that would render the evolved form less competitive to survive in nature, as is the case with regard to the concept of domesticated forms of the coconut. Such a selection would reduce the competitive ability of the evolved form to survive in nature. Survival under cultivation is a different story altogether.

5.5 The Nature and Characteristics of Present-Day Coconuts

A reading of the descriptions of present-day coconut cultivars as given in the International Coconut Genetic Resources Network (COGENT) publication, Conserved Germplasm Catalog (2010) shows that all the (so-called) present-day varieties/morphotypes show significant range of variation in all their morphological and physiological characters. This is also evident in the catalog of coconut germplasm access brought out by Rathnambal et al. (1995, 2000) (Tables 3.2 and 8.9). Present-day coconuts also show considerable variations in all their agronomical characters such as in the extent of seed setting, rate of seed germination consequent on the prevalence of seed dormancy, differences in growth rate, etc. Such variation is apparent, but possibly to a less extent also in the present-day landraces/varieties/morphotypes (Fig. 3.3). Part of the information given in Rathnambal et al.'s (1995) has been reworked and is presented in Table 3.2.

Such variability is considered typical of wild and weedy plants (Baker & Stebbins, 1965; Harlan, 1971, 1992; Harlan & De Wet, 1965, 1971; Nayar, 1973; and so on). This leads us to propose that present-day coconuts are basically wild in nature with incipient domestication taking place in just a few regions of the world. The terminology for the state of present-day coconuts as "semi-domesticated" was used first by Sauer (1971) in the context of the nature of coconuts in the Seychelles Islands. Nevertheless, he did not elaborate it further. He had merely stated it so in passing. Even though the dominant tall forms of coconut are predominantly cross-fertilizing in nature, if not completely, it has been seen that the pollination system in the coconut is opportunistic in nature, with varieties, seasons, locations, and even wind velocities, inducing variations in the pollinating system of the coconut to adapt the morphotype to local conditions. In the regions/countries of the world where the primary use of coconut is as a food item, varieties possessing higher yields in nuts, and more copra and oil appear to have been selected. However, this characteristic does not appear much in evidence in the Pacific Ocean island varieties. This needs to be studied further and confirmed.

Most of the present-day coconut landraces/morphotypes seem to possess varying levels of "ennobled" characters. At the same time, some of the characters do not appear to show any recognizable changes. Absence (as in *Nypa*) or even reduction in seed dormancy period is one such character (cf. Niral, 2012). There does not appear to have been a reduction in the period of dormancy in the present-day varieties as compared to the more primitive forms. Only the dwarfs have short seed dormancy. All the present-day tall cultivars show staggered germination. They take 60–240 days and even more time to germinate irrespective of the landrace/variety. A few nuts in some cases may take even longer than 240 days. A dry husk appears to be a prerequisite to initiate the germination process. This staggered manner of seed germination is a typical survival strategy characteristic of both wild plants and weeds to provide better chance for the survival of the organism in nature. Detailed studies of Jack Harlan and his associates (de Wit, Stemler, etc.) have confirmed this (cf. Harlan, 1992; and earlier papers).

Gunn et al. (2011) had proposed that the dwarfs are a domesticated form because they possess largely round fruits and self-fertilizing system of pollination. This is not borne out by available data. Years ago, Swaminathan and Nambiar (1961) too had made the same suggestion, but the generally prevalent opinion is that dwarfs are naturally occurring mutants in the coconut (Ninan & Raveendranath, 1983). This is supported by the presence of dwarfs in low proportions (<1%) in naturally occurring populations of coconuts. Further, the self-fertilization as prevalent in the dwarf forms is considered derived from cross-fertilization (which is prevalent in tall forms) (Stebbins, 1950).

5.6 Inferences

A present-day tall coconut palm produces 10–50 female flowers in a spadix and 10–14 spadices in an year in the tropics. In the germplasm collection maintained in Central Plantation Crops Research Institute (CPCRI) Kasaragod (Kerala), the percent bunch set varies widely from 40% to 97% (average: 67.3%) (Rathnambal et al.,1995; Tables 3.2 and 8.8). In the locally cultivated forms, the percent bunch set under average conditions will be much higher. In practice, 1–2 inflorescences in an year may turn out to be completely sterile (Patel, 1938).

In a similar way, the fruit set varied from 11% to 47% (average, 41.1% in talls, 31.5% in dwarfs; Table 8.8). When calculated on the basis of "buttons"

Table 8.8 Time of differentiation of Palmae Taxa (all in Mya: Million years ago)

	Gunn (2004)	Meerow et al. (2009, 2014)	Baker and Couvreur (2013a, 2013b)
Tribe Cocoseae (1/13)	53.26–61.09	63.8 (crown size)	6626/3.8–/54.43/55.77
Spiny/nonspiny divergence (Bactridinae/Attaleinae,Elaeidinae)	60.45–50.00		
Nonspiny clade (Attaleinae, Elaeidinae)	51.02–42.93		
Cocos Alliance (Cocos, Elaeis, Bactris, Butia Jubaea, Jubaeopsis, Syagrus, Parajubaea, Allagoptere, Attalea)	37.64–34.01		
Subtribe Attaleinae	–	43.7(27.2–50.7) 37.8 (crown size)	63.8/37.84–35.77/36.17
Beccariophoenix	–	–	30.89
Jubaeopsis caffra	35.72–37.68	28.5(16.6–41.6)	21.81
Voaniola	–	–	20.81
Allagoptera	–	20.4(9.9–23.4)	22.24/16.18
Attalea	40.0	13.0(8.6–20.5)	23.61/12.82
Butia	–	8.1(45–10.9)	14.71/7.89
Cocos nucifera	22.20–26.71	30.4(23.9–44.4)	23.61/3.92
Jubaea	23	14.5 (8.2–20.6)	14.71/NA
Lytocaryum	–	Butia/Jubaea 27.0(54.4–30.5) Syagrus (Lytocaryum embedded)	17.09/6.47
Syagrus	–	Cocos/Syagrus/Lytocaryum 34.9 (20.7–39.5)	27.6/19.98

Parajubaea	29.87–29.44 (P. torallyi)	Parajubaea/ Butia/jubaea/Allagoptera 31.6(17.4–34.2)	22.14/3.83
Subtribe Bactridinae			56.21/36.06
Acrocomia	—		33.04/16.62
Astrocaryum	—		33.04/28.09
Aiphanes	—		34.80/14.71
Bactris	—	—	33.00/20.68
Desmoncus	—	33.04–16.62	14.91/6.81
Subtribe Elaeidinae	—	33.04(stem)/16.62/crown	56.21/23.74
Barcella	—	—	23.74
Elaeis	—	34.80(stem)/14.71/crown	23.74/7.38

Table 8.9 Characteristics of present-day coconuts

Sl. no.	Character	Range	Average	Data source
1	Fruit set (%)			
	Talls	11–47	41.1	Table 3.2
	Dwarfs	25–38	31.5	
2	Bunch set (%)			
	Talls	40–97	67.3	Table 3.2
	Dwarfs	24–60	42.8	
3	Fruit length (cm)			
	Talls	20–39	30.2	Table 3.2
	Dwarfs	22–29	25.2	
4	Fruit length/breadth ratio			
	Talls	1.2–2.5		
	Dwarfs	1.3–1.9		
5	Tender nut water (ml)			
	Talls	75–890	301.5	Table 3.2
	Dwarfs	190–351	263.0	
6	Husk weight/fruit weight (%)			
	Talls	29.6–55.4	39.7	Table 3.2
	Dwarfs	30.5–58.5	40.1	
7	Kernel weight (g)			
	Talls	179–650	334.2	Table 3.2
	Dwarfs	121–290	200.0	
8	Shell weight (g)			
	Talls	100–285	155.7	Table 3.2
	Dwarfs	62–123	93.7	
9	Nut weight (g)			
	Talls	292–1389	649.8	Table 3.2
	Dwarfs	190–558	262.7	

(female flowers), the extensive observations taken by Patel (1938) showed that only less than 20% of the buttons developed into mature nuts (i.e., botanically "fruits"), 5% buttons shed before they opened, and 70–75% buttons shed within the first eight weeks of flower opening. Menon and Pandalai (1958) had also obtained similar results in extensive observations taken on more than 1000 palms for 25 years without interruption.

The states of high levels of sterility and low levels of fruit setting, as are prevalent in the coconut, are a typical characteristic of wild species. This is again best illustrated in the case of present-day cereals like rice. Incidentally, this characteristic may also be incidentally reflected in the coconut crop setting showing higher levels in some countries such as the West Indies, some states of India, and so on (Table 3.2). These are, of course, only very tentative pointers. They can be confirmed only after conducting more analyses.

The above inferences lend further weight to the observation that the present-day coconut is wild in nature in most parts of the world. In highly nobilized crop plants like wheat, rice, pulses, etc., we can observe that after 4 to 8×10^3 years/generations of selection and domestication, the seed-setting percentage is usually more than 95%.

The detailed analyses of 72 accessions of the coconut from the global collections of coconut genetic resources maintained at the CPCRI, Kasaragod (Kerala, India) (Rathnambal et al., 1995), gives some "feeble" pointers to this proposition. The wider prevalence of larger-sized nuts/fruits in the Philippines, smaller-sized nuts in the Laccadives, the data from West Indies and Sri Lanka all strengthen our suggestion that some selection of higher yields too have been underway here.

The present-day coconuts possess several other characteristics of wild/semiwild plants—the low seed set under natural conditions, staggered seed germination of even established morphotypes, an opportunistic pollination system, the significant levels of female sterility (button shedding, entire inflorescence sterility, staggered female flower receptivity even within an inflorescence), and so on (Tables 3.2 and 8.8). Several authors (Baker & Stebbins, 1965; Harlan, 1971, 1992; Harlan & De Wet, 1965, 1971; Simmonds, 1974; Stebbins, 1950; and several others) have dealt with the nature and properties of the progression from wild to domesticated forms. Thus, it seems appropriate to conclude that the present-day coconut is at best semiwild in nature.

CHAPTER 9

Spread

1. INTRODUCTION

Presently, the coconut is grown, or it grows naturally throughout the tropics. The Food and Agriculture Organization of the United Nations (FAO) (2011) has reported area–production statistics for 93 out of the total 243 political entities of the world (Chapter 1). It is now found on almost all of the islands and atolls of the Pacific and Indian Oceans, and along the coastal lowlands of tropical South and Southeast Asia, within c. 50 km of the seacoast, and, to a limited extent, in tropical Africa (Fig. 1.01). Ideally, it grows in the tropical lowlands, if suitable edaphic and climatic conditions are available (i.e., 1200–1500 mm of reasonably distributed annual rainfall and suitable soil).

2. FOR AND AGAINST NATURAL DISPERSALS

We have seen that the coconut has been part of the natural vegetation in much of the Indian Ocean and western and central Pacific islands even before the present-day humans had set foot there. It is also believed to have dispersed naturally through ocean currents to some of the nearby islands and atolls from regions or islands where the coconut was present as a component of the natural vegetation. Some dispersals have also been carried out by humans, as present-day humans began to spread out of their native homelands of the lower Hwang Ho River in middle-eastern China 50,000–70,000 ya to Island Southeast Asia and beyond to the Pacific Ocean islands. In their subsequent travels and voyages to these regions, they used to carry coconut fruits with them during their sailings for both water and food. This is evidenced from the findings of coconut-shell pieces and husks in former cave dwellings and archeological sites in these regions (Bellwood, 2013).

From the late 19th century through most of the 20th, strong opinions used to be expressed both for and against coconut dispersals by humans vs. natural means in the Indian and Pacific Oceans. We have already discussed this aspect in this volume. It is now generally accepted that natural dispersals by ocean currents would indeed have happened, but to a limited extent

The Coconut
ISBN 978-0-12-809778-6
http://dx.doi.org/10.1016/B978-0-12-809778-6.00009-7

only to nearby islands, but many of the dispersals, including all the long-distance ones, have been carried out deliberately by the humans.

Bates (1956) had observed that coconut is more dependent on man for dissemination. He added, "The coconut now seems perfectly capable of persisting as a part of strand vegetation with no interference by man. But, man does so frequently interfere that it is difficult to be sure what would happen to coconuts if their interference is erased."

Buck (1938), a former director of the Bernice P. Bishop Museum, Hawaii, United States, and in his time, an authority on Polynesia, had made similar observations. He stated that in uninhabited atolls, "coconuts had been planted by transient visitors…" When there is no evidence one way or another, it is difficult to rule out the possibility of planting.

Buck had made another observation that is relevant to our discussion. According to him, the Polynesians spread the coconut across Polynesia. When the first humans arrived in Polynesia, they had only a few plants available for food—only purslane (*Portulaca* sp.), roots of *Boerhavia*, seaweed, and possibly *Pandanus*. However, "when the first Europeans came, they had coconut, breadfruit, banana, taro, arrowroot, and sweet potato." This implies that it was the Polynesians who had spread coconuts and other crop plants.

Not yet explained in this context is that on most of the Indian and much of the western and central Pacific Ocean islands (and as remote as the Cook Islands 21°00′S, 160°00′W), the coconut, *Cocos nucifera* L., was already part of the natural vegetation when present-day humans first arrived there (cf. Kirch, 2000; Bellwood, 2013).

3. FROM THE OLD WORLD TO THE NEW WORLD

During the historical period, the spread of the coconut began as a secondary activity in the post-Middle Ages, when the European powers—initially, the Spanish, and then the Portuguese and English—wanted to break the monopoly of the lucrative Arab and Turkish spice traders from Asia to Europe, which was then being carried out by overland routes. In their endeavors to reach the Far East and India, the Portuguese took the eastern route around Africa and the Spanish the western route across the Atlantic Sea. These led to the landing of Portuguese explorer, Vasco da Gama, in Calicut, India, in CE 1498, via the Cape of Good Hope; the Spanish explorer, Christopher Columbus, in CE 1492 in Hispaniola (West Indies) across the Atlantic Ocean; and Vasco Núñez de Balboa the following year in the Pacific via the Drake Passage (Cape Horn). The rest is history. However, what is not widely appreciated is that introduction and spread of several cultivated plants from one part of the world to another was one of the great services rendered by these two countries, especially by the Portuguese. This also included the coconut.

During da Gama's first voyage to India (1947–49), he had stopped at Cape Verde Islands (16°10′N, 24°00′W) off West Africa (Fig. 9.1), Mombasa and Malindi (both in Kenya, 1°0′N, 37°18′E), and Goa and Calicut (Malabar–Konkan coast) in western India, among the places relevant to our discussion here. Both Kenya and the Malabar–Konkan coast had coconut cultivation at the time of da Gama's visit, especially the Malabar–Konkan region.

Cape Verde is a group of 10 volcanic islands located 570 km off West Africa (west of Dakar, Senegal). It is 4000 km² in area. It has a near-desert climate, because it receives only 265 mm precipitation annually. These islands are not suited for coconut cultivation, but the coconut can be grown here under protected conditions (Wikipedia, 2015).

During the late 15th century, these islands were beginning to come into prominence as a staging post for the then-increasing slave trade. After Columbus landed in Hispaniola in CE 1492 and Cabral in Brazil in CE 1500, a great demand had developed for slaves for establishing new colonies on the east coasts of North and South America. Incidentally, the Portuguese nobleman, Pedro Alvares Cabral, had conducted an expedition to India in CE 1500–02. He had taken a different route to India than Vasco da Gama, in that he traveled slightly southwestward, landing first at Porto Seguro (Bahia, Brazil; 16°26′S, 39°05′W), and in the process, discovering South America, which he then declared a Portuguese territory. From there, he traveled to India before returning to Portugal. In India, Cabral had visited all the ports that da Gama had visited earlier.

Harries (1977) proposed that the Cape Verde Islands had served also as the staging post for the dispersal of the coconut to the Western Hemisphere. Both Vasco da Gama and Pedro Cabral had anchored at Santiago, the main island, before returning to Lisbon. Da Gama's brother, who had accompanied Gama on his first voyage to India, fell ill while the ships berthed in Cape Verde and died there. This staging of ships from Portugal to the Cape Verde Islands continued for about 50 years from 1550 (Harries, 1977). He was, however, unsure who between da Gama and Cabral had introduced coconuts to the Cape Verde Islands. As indicated before, though the islands suffer from quasixeric conditions, it is possible to raise coconuts with supply of water. Coconut is one of the crops raised in this manner in the islands even today.

Harries (1977) does not actually provide any evidence for the introduction of the coconut in the Cape Verde Islands, from either India or East Africa, and its further spread to the New World. Nevertheless, he has provided some

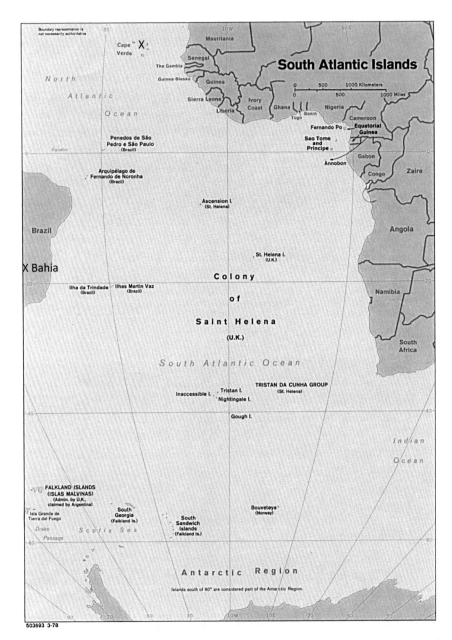

Figure 9.1 South Atlantic Ocean showing places mentioned in the text. *Reproduced with permission of the University of Texas Library.*

circumstantial and inferential evidence in support of this contention. All the authors have since cited this reference in the context of the coconut's spread into the New World. Harries appeared certain that the coconut was brought to Cape Verde by ship around the Cape of Good Hope from East Africa, and not overland across the central African tropical forests.

At the same time, there have been suggestions that the Asian tropical crops such as rice, banana, and others were being grown in tropical West Africa, and that they had been introduced there overland from East Africa across the tropical rain forests. Several oriental influences are evident in West Africa. This needs to be further investigated (cf. Jones, 1959; Hawley, 2008; for more details) vis-á-vis the transfer of the coconut from East Africa to tropical West Africa (cf. also Vansina, 1990).

In the literature, the credit for introducing coconuts to the New World is given to Vasco da Gama. However, E. D. Merrill (1954) had indicated that most of the dispersal of economic plants from Brazil to India and beyond appeared to have taken place from Bahia (Brazil), the route taken initially by Cabral.

4. COCONUTS ON THE PACIFIC COAST OF AMERICA

A "raging" controversy and discussion on when, where, and how coconuts arrived on the Pacific coast of the Neotropics continues to this day. It is now generally accepted that all of the remote ancestral lines of *Cocos* had originated in South America (Baker and Couvreur, 2013a,b; Meerow et al., 2014). It is, however, possible that the immediate ancestral lines of the coconut were likely present in the western Pacific and adjacent Indian Ocean regions to its west. Their biogeographical studies have shown that numerous dispersal events had taken place during the Miocene from South America to the Pacific and Indian Ocean regions.

The controversy about the presence of fully grown coconut stands on the Atlantic coast arose following the accounts of several early Spanish travelers and chroniclers reporting the presence of coconuts along the west coast of the Neotropics not long after the landing of Columbus in Hispaniola in CE 1492. It is now generally agreed that practically all such reports were not of the coconuts, but of other palm species that possessed broadly gross morphological similarities to the coconut. This leaves us still with a couple of reports on the presence of coconuts on the west coast of Panama at the time of Columbus' landing in America. Thus, the question that we raised previously remains.

Dennis and Gunn (1971) reviewed the past literature on the subject. Then, they concluded that natural dispersal of coconuts by ocean currents in either direction across the Pacific was highly improbable. In this context, Ward and Brookfield (1972) used a computer simulation model taking into account wind and ocean currents' velocity and similar parameters to see if drift voyaging of coconuts could have brought them from the west Pacific to the American coast.

The first report of coconuts in America, according to Ward and Brookfield, was of Peter Martyr in 1516 from the Coiba expedition of Gonzalo de Badajoz in CE 1514. "On the South Sea, there are various islands to the west of the Gulf of San Miguel and Isla de Rey Rica (Pearl Island, 8°27'N 75°55'W) in the greater part of which there grow and are cultivated trees that produce the same fruit as in the country of Colocut (Calicut), Cochine (Cochin), and Camemore (Cannanore), where the Portuguese have their spice markets" (quoted in Sauer, 1967). Martyr reported that "the germs" of the trees were brought by the waves from Burma (Bruman, 1944, p. 224, 1945, 1947; Ward & Brookfield, 1972). Ward and Brookfield (1972) concluded that it was extremely unlikely that coconuts could have crossed the central Pacific in either direction as unaided voyagers while remaining viable, because this region appeared to be nautically difficult to cross in medieval times.

The studies of the Panama Tall coconut variety by Lebraun et al. (1998) using Restriction Fragment Length Polymorphism (RFLP), Lebraun et al. (2005) using markers, and Baudouin, Lebron, Berger, Myrie, and Been (2008) with microsatellites, showed that Panama Tall possessed fruit characters similar to those of San Ramon and Tagnanan varieties of the Philippines. In a more detailed study, Baudouin and Lebrun (2009) studied 1215 palms to determine the relationships among the Pacific Ocean varieties using 30 microsatellites. The material also included 104 Panama Talls. The 1215 palms grouped themselves into 11 populations, all from Southeast Asia and the Philippines. Further, based on the similarities of some pottery objects found in Ecuador (Bahia de Caraquez) with those of Southeast Asia, they proposed that early navigators from Southeast Asia first brought coconuts to Ecuador in 2250 BP (Baudouin et al., 2006).

Baudouin, Gunn, and Olsen (2014) examined the writings of the Spanish Chronicler, Oviedo (Gonzalo Fernandez de Oviedo, 1478–1557). Oviedo had used fruit and seed size to distinguish coconuts in a coastal Panama population, in which tidal dispersal of coconuts has apparently been taking place. The authors' genetic analyses indicated that it was apparently brought from the Philippines. Baudouin et al. (2014) then gave details of other locations on the western American coast, where Oviedo had recorded the presence of coconuts. One such location was close to the town of Aguadulce (Panama), near Santa Maria River (8°9′N, 80°31′W). This material was attributed to a sample of Philippines Tall collected by one J. L. Renard. The other pre-Columbian populations came from the Pacific coast of Panama, Cost Rica, and Peru. They suggested that all of them originated from the nuts of only two to five palms each, and that all of them had initially originated in the Philippines. Baudouin et al. (2014) also criticized the observations made by Harries (2012) and Clement et al. (2013) about the presence of the coconut in the Americas at the time of Columbus landing in America, because they had not also supported their observations with adequate evidences.

Harries (2012) elaborated the suggestion first made by him in 1978 that the coconut was introduced to the Americas by the Spaniards during the period, CE 1565–1815, while returning from the Philippines using the Manila–Acapulco galleon route. The author had given some details surrounding the story, but it had not included any evidence supporting this proposition.

The Acapulco–Manila galleon line was a third route by which Mexican economic plants and weeds had reached the Orient (Merrill, 1954). The first galleon crossed the Pacific in 1565 and the last in 1815. This was a Spanish government-sponsored and controlled shipping line. Usually, only one ship was permitted each way per year, but, occasionally, two or even three, and also some clandestine private enterprise interlopers, used to ply this route. The usual time to travel westward was 3 months, and sometimes, it extended up to 5 or 6 months. The eastward voyage used to take 5–7 months. Each manifest used to carry up to 2000 tons of valuable goods and hundreds of passengers. However, specific information on the transport of coconut was lacking. In addition, the eastward voyaging time of 5–7 months could have had some adverse bearing on the viability of coconuts, in case they were being brought from the Philippines to western Mexico. Although a strong possibility exists of this route having been used, there is no confirmatory evidence yet in this regard.

Piperno and Pearsall (1998) studied the archeology of the Panama coast. They did not observe any coconut remains, but recovered broken endocarps of several other *Attalea* subtribe (Cocoseae) palms, such as those of *Acrocomia acauleata*, *Attalea butyracea*, *Bactris major*, *Elaeis oleifera*, and *Astrocaryum* sp. Among these, the fruits of *Attalea butyracea* "look somewhat like dried coconuts". The fruits measured 4.5–8.5 cm × 3.0–4.5 cm in size. The broken endocarps of *Attalea butyracea* were dated to 7000 BP.

Clement et al. (2013) reviewed the introduction and spread of coconuts in the North and South American continents. They attributed to Bruman (1944) that coconuts from the Cape Verde Islands were planted in Bahia, Brazil, which, according to them, were the first coconuts planted in the Americas. They stated further that in 1569, one Alisara de Mendano introduced coconuts from South Sea Islands to Colima on the Pacific coast of Mexico. However, in Bruman (1945), he had stated that it was actually one Herman Cortes, who had brought two dozen ripe coconuts from Panama before 1539 and planted them in Colima.

Clement et al. had examined also the languages of the local language groups in the Panama region, albeit in a preliminary manner. They could not find any word for the coconut in any of the local dialects/languages. Based on their review, they concluded that "a pre-Columbian presence of coconuts in Panama and surrounding areas is not attested by paleo-bio-linguistic [sic] evidence. Indeed, the available evidence strongly suggests that the modern occurrence of the plant in the region is accountable (attributable) to European introduction in historic times.... Coconut arrived immediately before European conquest, rather than in 2250 BP, as suggested by Baudoin and Lebrun (2009), or, it arrived after the European conquest. Either way, the genetic evidences and historic records need further study."

Thus in a way, the controversy continues to simmer about the pre-Columbian presence of the coconut on the west coast of the American continents. Indeed, all the biological evidence available from molecular studies, the limited morphological evidence available from the comparison of Panama Tall with the West Indies tall forms (Whitehead, 1965), and the reasonably credible travel accounts of some early explorers such as Oviedo do indicate the presence of the coconut palm in a few isolated and scattered locations on the Pacific coast of tropical America.

As mentioned earlier in this text, that in the absence of hard evidence for deliberate introduction of the coconut from the western Pacific Ocean to the western American coasts, we assume that some accidental landfalls of Polynesian canoes could also have introduced coconuts to the Americas. The pre-Columbian recovery of chicken bone from Arauco Peninsula, Chile, dated to CE 1304–1424 has already been cited (Storey et al., 2007). We have seen also that naturally occurring coconuts have been recorded as far as the Cook Islands (east Polynesia) before the arrival of modern humans. (Matisoo-Smith, 2007; Storey, Quiroz, & Matisso-Smith, 2011; Storey et al., 2007) It is also possible that naturally occurring coconuts were present in French Polynesia to the east of the Cook Islands. This is apparent from the writings of Peter Martyr (1516) [quoted by Ward and Brookfield (1972)]. From French Polynesia, the distance to the tropical coasts of western tropical America is only about 5000 km. (Zizumbo-Villareal, 1996)

Notwithstanding the digressions in the details, all the above observations by various authors give good support for the presence of the coconut on the American continent before the arrival of Columbus in AD 1497. The time of arrival of the first coconuts from the Philippines continues to be a moot question.

5. SPREAD OF THE COCONUT IN SOUTH AMERICA

We have only a very sketchy information about the spread of the coconut in the New World. The following account has been prepared from the documentation, coconut line, kindly supplied by Hugh Harries (2015).

The coconut was introduced into Puerto Rico in 1549 from the Cape Verde Islands (Harries, 1978). In 1915, seed nuts from the Pacific were introduced to Jamaica and elsewhere after the Suez Canal was opened. Small (1929) recorded that one Jean Baptiste Christophe Fusee Aublet, who had visited Cayenne (French Guiana) during 1762–64, had observed that missionaries had earlier introduced the coconut there. One Francis Moore in 1974 had mentioned about a garden of West Indian plants raised in Savannah, Georgia (32°5′N 81°6′W), overlooking the Savannah River (Small, 1929). However, we know that this was unlikely to have survived for long because of the extreme cold prevailing in the area during winter. Presently, coconuts are known to be growing in the continental United States only in Florida and that also for only ornamental purposes.

One Capt. Nathanial Uring in 1726 recorded seeing a lone coconut palm on the Honduras coast between Cabo de Gracias a Dios and Cabo Camaron on the Atlantic coast (15°N, 85°W approx.), after his ship had wrecked there. He felt that it might have arisen from a drift nut (Clement et al., 2013).

Wafter has given a detailed account of his visit to Cocos Island (Costa Rica) in 1685. He had found that the coconuts were growing luxuriously there. In Brazil, which is presently one of the major coconut-growing countries of the New World, William Piso, who had spent 7 years in Brazil, had compiled a list of palms growing in that country, and this included the coconut (Small, 1929).

Incidentally, the well-known botanist, Merrill (1945), while commenting on Cabral's landing in Recife, Brazil, in AD 1500 had observed that it was through this port that the Portuguese had carried out most of the economic plant exchanges with the rest of the world. Francisco Hernandez, who had been sent on a mission to Mexico reported in 1575 about seeing coconuts growing on the west coast of tropical Mexico. (Zizumbo-Villareal, 1996)

Thus, we have only sketchy information about the introduction of the coconut into the New World and its further spread to the other regions of South America. Many of the earlier reports of sighting coconuts by the early Spanish travelers were in reality of other palm species that had broad morphological similarities with the coconut palm. Besides, most of these early Spanish and Portuguese travelers might not have been sufficiently knowledgable about the coconut palm and its morphological features.

Incidentally, the history of coconuts in Africa is not much better than this. At present, we have only approximate accounts of its introduction even on the east coast of Africa including Madagascar.

CHAPTER 10

Afterword: Does the Coconut Have a Future?

1. INTRODUCTION

In the last nine chapters, we had discussed some fundamental aspects about the coconut. We also discussed the rise and fall of the coconut, the uncertain aspects about its remote origin, and the apparently primeval nature of the present cultivated forms. We are ending the volume with a question mark on the future of the coconut.

The coconut oil was the leading vegetable oil traded in the world for nearly a century from the mid-19th century. In 1961, when the Food and Agriculture Organization (FAO) began publishing area–production statistics of cultivated plants, the position of coconut oil #4 among the 13 vegetable oils in the world, and in CE 2011, it had been relegated to ninth position among them (Table 1.1). We discussed earlier the reasons for this apparent fall of coconut oil.

However, of late, the coconut is "re-emerging as a sexy new superfood," to quote from Harvard Health Publication (2014). This appears to be because of two main factors. One is the increasing realization that "even though coconut oil has about the highest percentage of saturated fats—84%—it is composed largely of medium chain fatty acids, and the liver coverts these smaller molecules into energy more easily, so they are less likely to form artery-clogging LDL. ... Compared with a diet in refined carbohydrates, it is a better source of calories than white flour... Further, coconut oil gains a slight edge over other vegetable oils in baking, because it remains solid at room temperature. For this reason, more companies are using coconut oil in baking, to replace hydrogenated soybean oil in their products" (Harvard Health Online, 2014). The American Heart Association continues to name the coconut and oil-palm oils to be avoided along with all the animal fats and butter.

Despite the above properties, the native systems of medicine in the various cultures of South and Southeast Asia, Oceania, and the Indian Ocean islands ascribe several benefits to the coconut oil, such as promoting weight loss, strengthening immunity, producing a sobering effect on diabetes, and an excellent emollient even for newborn babies, and so on.

The Coconut
ISBN 978-0-12-809778-6
http://dx.doi.org/10.1016/B978-0-12-809778-6.00010-3

At the same time, several other aspects continue to cause anxieties about the future of the coconut. The foremost among them is that more than half of the present-day coconut palms, growing particularly in the Indian and Pacific Ocean islands, are in a senile condition and require replacement. Even in continental South and Southeast Asia such as Sri Lanka, the Philippines, India, and Malaysia, where the coconut continues to have an important place in their respective economies and people's livelihoods, only a modest level of replanting is taking place. In Indonesia, newer areas are being brought under cultivation, but its pace has come down substantially. Lack of national and international funding support for planting and the uneconomic commodity prices appear to be the main reason for this deplorable state of affairs.

Today, the coconut is an orphan crop plant. In the 1970s and 1980s, there were some fledgling initiatives from the World Bank and FAO to promote the crop, but it did not go beyond a few uncertain steps. These efforts had premature closure.

A couple of studies conducted in the recent past highlight the difficult situation facing the coconut industry.

The Philippines is the world's largest production of coconuts accounting for 27% of global production (Table 1.7). In the Philippines, the coconut covers 30% of the farmland, which is even higher than that of its most important crop, rice. It contributes nearly half of the old agricultural exports of the country. Yet the crop generates only the lowest farm value per hectare among all the crops of the country. This was attributable to low yields, monocropping, and underutilization of land under the crop. This study promoted by the United States Agency for International Development (USAID) showed also that the coconut industrial sector spanned many activities, but they were characterized by overcapacity consequent on stagnation in farm production.

The reasons for this state of affairs have been ascribed to mainly inadequate commitment at the top decision-making level to redeem the situation and the shortage of long-term financing needed for development of this perennial crop (Dy & Reyes, 2008).

In 2007, the Australian Center for International Agricultural Research (ACIAR) commissioned a report to develop a strategic basis for the future research and development (R&D) investments in the coconut in the Asia-Pacific region. This region accounts for as much as 85% of the coconut production in the world. In much of the area, according to the study "coconuts play a pivotal rote in the livelihood of quasi-subsistence small holders, who dominate primary production in the region, and as a key source of cash income and nutation." The study showed also that a "significant proportion of the (coconut) stands is used to provide shade for low-input production of other food and cash crops" (Warner & Querke, 2007).

The report estimated that 75% of the palm population was senile in the region. However, the smallholders have been apparently reluctant to take up replanting programs for two reasons—the loss of production of coconuts and intercropped crops during the extended juvenile phase that replanting entails, and second, the reluctance of the smallholders to manage the risk because of the limited incentives that are available to do this. The study pointed out that marketing of virgin coconut oil and the potential to use coconut oil as a replacement for diesel in internal combustion engines might offer only limited opportunities.

The report then went on to suggest a multipronged research strategy to face the situation (Australian Center for International Agricultural Research (ACIAR), 2006).

Some of the major coconut-growing countries—the Philippines, India, Sri Lanka, Kenya, for instance—have government-supported national bodies such as the Philippine Coconut Industry, Coconut Development Board (of India), Coconut Development Authority of Sri Lanka, and Kenya Coconut Development Authority—to oversee the development of the coconut. Some of them have been in existence for more than 70 years. But overall, their effectiveness and impact have been limited.

The Asian and Pacific Coconut Community (APCC), Jakarta, Indonesia, is an intergovernmental organization established in 1969 under the aegis of United Nations Ecocomic and Social Commission for Asia and the Pacific (UN-ESCAP) with 18 members of the region "to promote, cocordinate [sic], and harmonize all the activities of the coconut industry, which sustains the lives of millions of small farmers as well as those engaged in production, processing, and marketing of coconut products." This entailed a view to improve the socioeconomic conditions of all stakeholders of the coconut community in the member countries, particularly the small coconut farmers. However, possibly because of its financial limitations, it cannot be stated that the APCC has been able to make any waves toward achieving the objectives or its mission or vision.

Thus, the coconut continues to remain an orphan commodity. At the same time, for most of the populations of the hundreds of inhabited islands of the Indian and Pacific Oceans, the coconut continues to be a secondary staple crop. Their lives and livelihoods are going to be in jeopardy in the next 20–30 years, when the last of the coconuts that had been planted by the plantation industry would cease to be productive and the twin effects of climate change—sea level rise, and temperature and weather uncertainties— would begin to be felt in these regions.

The only way out of this situation appears to be to recognize the coconut as a mandated crop of the International Center for Research on Agroforestry (ICRAF), Nairobi, East Africa, which functions under the Consultative Group on International Agricultural Research (CGIAR).

It is not widely appreciated that the coconut is by far the most appropriate agroforestry species for lowland tropics. Though every part of the coconut tree is used traditionally for various purposes, at least five parts—the wood, coconut shell, pith, fiber, and kernel—find active industrial, agricultural, and food-industry applications. The coconut palm can be considered the most appropriate tree species for intercropping, multistoried cropping, and composite farming—involving integration of agriculture, animal husbandry, including dairying, poultry rearing, piggery, and fisheries. This has been successfully demonstrated for several decades now for lowland tropical situations in Central Plantation Crops Research Institute (CPCRI), Kasaragod, Kerala, India (5 m msl, 3200 mm precipitation).

In a private communication, Vinay Chand (2016), an authority on coconut marketing, has estimated that the worldwide utilization of the nut, husk, and shell is as follows: Copra, 39%; Fresh consumption, 31%; Drink, 13%; Klentik oil [refined virgin coconut oil (VCO)], 7.3%; Desiccated coconut, 3.8%; Water, 2–3%; Coconut milk, 1.4%; Others, 1.8%; Coir, 17.5%; Activated charcoal, 8.8%.

The APCC has estimated the extent of international trade in the major coconut products (in 2010/11) as follows: Coconut oil, 1.86 Mt; VCO, 6000 t; Desiccated coconut, 331,000 t; Shell charcoal, 225,000 t; Activated charcoal, 110,000 t; Coir products, 630,000 t; and Coconut water, 17,000 t. The coconut production during 2011 was 2.99 Mt (Arancon, 2013).

The coconut industry appears to have done reasonably well despite not any significant official support in their efforts. With some official patronage, the outlook is likely to improve. Nevertheless, a makeover of the coconut is called for to enable the palm to face the newer emerging challenges.

2. R&D EFFORTS FOR TRANSFORMING THE COCONUT IDEOTYPE TO FACE THE NEWER CHALLENGES OF THE 21ST CENTURY

The most import task is to induce determinacy in the coconut. The average age of the coconut palm is 80–100 years. But, usually even under good managed conditions, the palm begins to yield low after 60–70 years of age.

Further, one of the major problems limiting nut yields in the coconut is the loss of yield resulting from the delayed harvesting of the coconut. Generally, in most of the continental regions of Asia and also Sri Lanka, the nut harvest is done by professional climbers. It is now becoming increasingly scarcer to get their services for harvesting coconuts. Elsewhere in the world, especially in the Indian and Pacific Ocean islands, professional climbers are not available, and the locals harvest the produce by collecting the fallen nuts from the ground. This procedure is estimated to cause more than 30% yield loss. In a few countries like Malaysia, coconuts are harvested using long poles fitted at one end with a sickle. This is the practice followed in oil-palm harvesting also. Now, advanced, lightweight, durable, telescopic poles made of carbon fiber are available for this purpose. They can reach a height of 20 m. With this length of the pole, it will be possible to harvest ripe coconuts standing on the ground using these pole harvesters. The present difficulties in coconut harvesting can be overcome if the coconut palm completes its productive life in 40–50 years. At this age, the palms would be 20–25 m tall, and it would be possible to harvest the produce using the above-mentioned telescopic poles.

DNA sequences are available in the National Genetic Sequence Data Base (GENBANK) that can impart determinancy in the coconut. We may transfer this gene to the coconut. Some examples are Zhang et al. (2003), Kwak et al. (2008), Tian et al. (2010), and Mir et al. (2014).

We shall examine one of these studies in some detail (Tian et al., 2010). Under cultivated conditions, soybean varieties may have either a determinate or an indeterminate life cycle. In cultivars with indeterminate growth, the growth gradually slows down and ceases as the pods begin to develop, and the plant increasingly diverts nutrients and water to the developing pods. The putative ancestor of soybean, *Glycine soja*, has an indeterminate growth habit. This character is monogenic in inheritance (*Dt 1/Dt1*).

Dt 1 is a homolog [designated as *Gm Tfl 1* of *Arabidopsis* terminal *flr1* (*Tfl 1*)], which is a regulatory gene encoding a signaling protein of the shoot meristem (Tian et al., 2010). The transition from indeterminate to determinate phenotypes appears to be associated with four single-nucleotide substitutions in the *Gm Tfl 1* gene, each leading to an amino acid change. This may have happened several millennia ago in soybeans consequent on selection by the first farmers. The authors, Tian et al. then confirmed this by suitable experimentation. The authors demonstrated that the *Gn Tfl1* gene complements the functions of *TFL 1* in *Arabidopsis*. However, the *Gm Tfl1* homeolog (in soybean) "appeared to be sub- or neofunctionalized." This was revealed by the differential expression of the two genes at multiple plant development stages and by allelic analyses of both loci.

In soybean, stem growth is controlled by the *Dt1* locus and indeterminate growth is expressed by *Dt 1/Dt 1*. This character may be expressed as fully or partially dominant, due to the operation of the other genes that are also responsible for this character, *Dt 2*, and possibly also *dt 1 -1*. The authors have fine-mapped the *Dt 1* locus as a major quantitative trait locus. It is located between the two simple sequence repeat markers, *Sa-999* and *Sat-229* on linkage group L (now chromosome 19). The authors have also indicated that *Gm Tfl 1* may be a candidate gene for *Dt 1*.

The DNA sequences for these should be available in the GENBANK. It is strongly urged that the scientists initiate a program for transforming the coconut into the determinate habit with a life span of 40–60 years using an appropriate DNA sequence for determinacy.

A second aspect that requires the attention of the scientists is to enhance the yield and other economic attributes of the coconut palm by further nobilizing it.

From the evolutionary point of view, today's coconut is semiwild in nature with the coconut in different regions of the world showing only low levels of ennoblement. One of the characteristics of present-day cultivated coconut is the high prevalence of floral sterility. Even established landraces in the CPCRI Kasaragod germplasm collection show only 24–97% floral bunches setting and 6–47% fruit setting in bunches (Table 3.3). From this, we can safely assume that in the local (unselected) populations in the various countries, the extent of bunch and fruit setting would be much lower than these values. With an accelerated and focused improvement program, it should be possible to mobilize these characters steadily. A perusal of Table 3.3 will show that some forms collected from regions/communities that give the coconut a prominent place in society, both the bunch and the fruit setting are higher. Compare the values for Zanzibar and the Philippines with those in Andaman islands, for instance.

Another aspect that needs attention is the prevalence of numerous maladies in the coconut with uncertain etiology in different regions of the world. This began with lethal yellowing and root (wilt) diseases, both of which have been festering the coconut for more than a century. Even now, no lasting solutions have been developed against them. In the meantime, new maladies continue to occur in other regions. The recent maladies reported from Mozambique and Ivory Coast are examples of this.

Except for isolated and ad hoc steps, no concerted efforts have been under way to manage these diseases and take up systematic programs to determine their etiology.

It is urged that a permanent body may be set up to develop a strong and durable program to confront and manage the situation.

We have seen that 50–80% of the palms the world over are in need of replanting. A concerted effort is called for to take up a global replanting program, say within 20 years. Financial guarantees for this should be sought from multilateral agencies. This will be facilitated if the coconut is formally accepted as a mandated crop of ICRAF. The coconut is presently an orphan crop. Historically, it had its rise and fall. A plan of action is necessary to re-establish this staff of life for the scores of islands, and this will also resurrect lowland coastal communities. Further details are given in Nayar (2007).

We shall conclude this chapter with a consideration of the origins and phylogeny of the coconut. Despite the importance of the coconut, there has only been a single study done on the phylogeny of the coconut, *Cocos* (Gunn, 2004). Of course, valuable additional information has emerged from the studies of Baker et al. (2009, 2011), Baker and Couvreur (2013a,b), and Meerow et al. (2009, 2014). Nevertheless, the results obtained by them are at variance with each other and variable even within their own studies regarding the genus *Cocos*.

One reason for this may be the nebulous state of generic delimitation of the unarmed Attaleinae genera. The recent proposal (Noblick & Meerow, 2015) to merge *Lytocaryum* with *Syagrus* may attest to this proposition. Further, the two genera showing more consistant sister relationships with the genus *Cocos*, *Attalea*, and *Syagrus*, are both relatively large genera with 73 and 31 species, respectively. All the studies done so far have used only small samples of these genera. Further, both the number of morphotypes of the coconut and the manner of their choice appear to have been done arbitrarily without giving proper consideration of their origins. Hence, a comprehensive study using all the species of *Attalea* and *Syagrus* and a wider choice of the cultivated coconut is called for.

Let us hope that the coconut will receive more attention, better support, and greater participation of the scientific community to look into the unsolved problems of this most useful multipurpose species, the coconut, in its centennial year of research. Then it will have a future.

REFERENCES

Abe, R., & Ohtani, K. (2013). An ethnobotanical study of medicinal plants and traditional therapies on Batan Island, the Philippines. *Journal of Ethnopharmacology, 145,* 554–565.

ACIAR. (2006). *Coconut revival: New posssibilities for the Tree of Life.* Canberra, Australia: ACIAR, 103 pp.

Adkins, S., Foale, M., & Harries, H. (2002). Growth and production of coconut. In *Encyclopedia of life-supporting systems. Soils, plant growth, and crop production* (Vol. III). Wiley. 31 pp.

Ahuja, S. C., Ahuja, S., & Ahuja, U. (2014). Coconut – history, uses, and folklore. *Asian Agri-History, 18,* 221–248.

Ali, J. R., & Aitchison, J. C. (2008). Gondwana to Asia: plate tectonics, palaeogeography, and biological connectivity of the Indian subcontinent from the Middle Jurassic through latest Eocene (166–35 Mya). *Earth Science Reviews, 88,* 145–166.

Allaby, M. (2006). *Dictionary of plant sciences.* New Delhi: Oxford University Press. 510 pp.

Allen, J., & Gosden, C. (Eds.). (1991). *Report of the Lapita homeland project. Occasional papers in prehistory #20.* Canberra, Australia: Australian National University.

Alzina, F. I. (1668). On the palms which are called *Cocos* and their great usefulness (L.B. Uichanco 1931, Trans.) *Philippine Agriculturist, 20,* 435–446.

APG III. (2009). An update of the Angiosperm Phylogeny Group classification for the orders and families of flowering plants. *Botanical Journal of the Linnean Society, 161,* 105–121.

Arancon, R. N. (2013). *Personal communication.*

Ashburner, G. R., Thompson, W. K., Halloran, G. M., & Foale, M. A. (1997). Fruit component analyses of south Pacific coconut palm populations. *Genetic Resources and Crop Evolution, 44,* 327–335.

Asmussen, C. B., & Chase, W. M. (2001). Coding and noncoding plastid DNA in palm systematics. *American Journal of Botany, 88,* 1103–1117.

Asmussen, C. B., Dransfield, J., Deickman, V., Barford, A. S., Pinaud, C., & Baker, W. J. (2006). A new subfamily classification of the palm family (Arecaceae): evidence from plastid DNA phylogeny. *Botanical Journal of the Linnean Society, 151,* 15–38.

Bailey, F. M. (1902). *The Queensland flora* (Vol. 5). Brisbane: Diddams. (cf. Dowe & Smith, 2002).

Baker, W. J., & Couvreur, T. L. P. (2013a). Global biogeography and diversification of palms sheds light on the evolution of tropical lineages. I. Historical biogeography. *Journal of Biogeography, 40,* 274–285.

Baker, W. J., & Couvreur, T. L. P. (2013b). Global biogeography and diversification of palm sheds light on the evolution of tropical lineage. II. Diversification history and origin of regional lineages. *Journal of Biogeography, 40,* 286–298.

Baker, W. J., Norup, M. V., Clarkson, J. J., Couvreur, T. L. P., Dowe, J. L., Lewis, C. E., et al. (2011). Phylogenetic relationships among arecoid palms (Arecaceae: Arecoideae). *Annals of Botany, 105,* 1417–1436.

Baker, W. J., Savolainen, V., Asmussen-Lange, C. B., Chase, M. W., Dransfield, J., Forest, F., et al. (2009). Complete generic-level phylogenetic analyses of palms (Arecaceae) with comparisons of supertree and supermatrix approaches. *Systematic Biology.* http://dx.doi.org/ 10.1093/syslro/sypo21.

Baker, H. G., & Stebbins, G. L. (Eds.). (1965). *Genetics of colonizing species.* New York: Academic Press.

Balakrishnan, N. P., & Nair, R. B. (1979). Wild populations of *Areca* and *Cocos* in Andaman & Nicobar Islands. *Indian Journal of Forestry, 2,* 350–363.

Balick, M. J. (1984). Ethnobotany of palms in the neotropics. In G. T. Prance, & J. A. Kallunki (Eds.), *Ethnobotany of the neotropics. Advances in ethnobotany* (pp. 9–23). Bronx, NY, USA: New York Botanical Garden.

Balick, M. J. (Ed.). (2009). *Ethnobotany of Pohnpei, Micronesia*. Honolulu, HI, USA: University of Hawaii Press. 606 pp.

Ballance, P. F., Gregory, M. R., & Gibson, G. W. (1981). Coconuts in Miocene turbidites in New Zealand: possible evidence for tsunami origin of some turbidity currents. *Geology*, *9*, 592–598.

Barrau, J. (1961). *Subsistence agriculture in Polynesia and Micronesia (Bull # 223)*. Honolulu, HI, USA: BP Bishop Museum, 94 pp.

Barrau, J. (1965). An essay on ethnobiological adaptation to contrastive environments in the Indo-Pacific area. *Journal of the Polynesian Society*, *74*, 329–346.

Bates, M. (1956). Man as an agent in the spread of organisms. In W. L. Thomers (Ed.), *Man's role in changing the face of the earth* (pp. 788–804). Chicago, IL, USA: University of Chicago Press. (cf. Wiens, 1962).

Batugal, P., Rao, V. R., & Oliver, J. (Eds.). (2005). *Coconut genetic resources*. Serdang, Selangor DE, Malaysia: IPGRI – APO. 797 pp.

Batuta, I. (1929). In H. A. R. Gibb (Ed.), *Ibn Batuta: Travels in Asia and Africa*. Delhi: Manohar Publishers. (Reprinted 2001).

Baudouin, L., Gunn, B. F., & Olsen, K. M. (2014). The presence of coconut in southern Panama in pre-Columbian times: clearing up the confusion. *Annals of Botany*, *113*, 1–5.

Baudouin, L., Lebron, P., Berger, A., Myrie, W., & Been, B. (2008). The Panama Tall and the Maypan hybrid coconut in Jamaica. *Euphytica*. http://dx.doi.org/10.1007/s10681-007-9568-2.

Baudouin, L., & Lebrun, P. (2009). Coconut DNA studies support the hypothesis of an ancient Austronesian migration from Southeast Asia to America. *Genetic Resources and Crop Evolution*, *56*, 257–262.

Baudouin, L., Lebrun, P., Konan, J. L., Ritter, E., Berger, A., & Billotte, N. (2006). QTL analysis of fruit components in the progeny of a Rennell Island Tall coconut (*Cocos nucifera* L.) individual. *Theoretical and Applied Genetics*, *112*, 258–268.

Beaglehole, J. C. (Ed.). (1962). *The Endeavour Journal of Joseph Banks, 1968–1971*. Sydney: Angus & Robertson. (cf. Dowe & Smith, 2002).

Beccari, O. (1916). It genera *Cocos* Linn. Ele Palmae affini. *L'Agricultura Coloniale*, *10*, 435–471 489–532, 585–623.

Beccari, U. (1917). The origin and dispersal of *Cocos nucifera*. *Philippine Journal of Science, Series C: Botany*, *12*, 27–43.

Bellwood, P. (1987). *The polynesians: Prehistory of an island people* (revised ed.). London: Thames and Hudson. 175 pp.

Bellwood, P. (2004). *First farmers: The origins of agricultural societies*. Malden, MA, USA: Blackwell Publishing. 360 pp.

Bellwood, P. (2013). *First migrants: Ancient migrations in global perspective*. Malden, MA, USA: Wiley Blackwell. 308 pp.

Bellwood, P., Fox, J. J., & Tryon, D. (Eds.). (1995). *The Austronesians*. Canberra, Australia: Australian National University.

Berry, E. W. (1914). *The upper cretaceous and eocene floras of South Carolina and Georgia*. US Geological Survey Professional Paper # 84. 5–200.

Berry, E. W. (1926). *Cocos* and *Phymatocaryon* in the Pliocene of New Zealand. *American Journal of Science, Series C: Botany*, *12*, 181–184.

BI. (2008). *Key access and utilization descriptors for coconut genetic resources*. http://www.biodiversity.international.org/publications/web – version/108/.

Black, R. (1996). A new coconut locality in Hawkes Bay. *Geological Society of New Zealand Newsletter*, *92*, 37–38.

Blench, P. (2007). New palaeozoological evidence for the settlement of Madagascar. *Azania*, *42*, 69–82.

Bourdeix, R., Johnson, V., Tuia, S. V. S., Kape, J., & Planes, S. (2013). Traditional conservation areas of coconut varieties and associated knowledge in Polynesian islands. In S. Larrue (Ed.), *Biodiversity and societies in the Pacific islands* (pp. 199–223). http://www.co gentnetwork org/images/publications/Bourderx et al–Biodiversity.

Bremer, K. (2000). Early Cretaceous lineages of monocot flowering plants. *Proceedings of the National Academy of Sciences of the United States of America, 97,* 4707–4711.

Bromham, L., & Penny, D. (2003). The modern molecular clock. *Nature Genetics, 4,* 216–224.

Bruman, H. J. (1944). Some observations on the early history of the coconut in the New World. *Acta Americana, 2,* 220–243.

Bruman, H. J. (1945). Early coconut culture in western Mexico. *American Historical Review, 25,* 212–223.

Bruman, H. J. (1947). A further note on coconuts in Colima. *Hispanic American Historical Review, 27,* 572–573.

Buck, P. H. (1938). *Vikings of the sunrise.* Philadelphia, PA, USA: J.B. Lippincott. (cf. Wiens, 1962).

Buckley, R., & Harries, H. C. (1984). Self-sown wild-type coconuts from Australia. *Biotropica, 16,* 148–151.

Buerki, S., Forest, F., Alvarez, N., Nylander, J. A. A., Arrigo, N., & Sanmartin, L. (2011). An evaluation of new parsimony-based vs. parameted inference methods in biogeography. *Journal of Biogeography, 38,* 531–550. (cf., Baker & Couvreur, 2013b).

Burkill, I. H. (1919). Some notes on the pollination of flowers in the Botanical Garden, Singapore and other parts of Malay Peninsula. *The Garden's Bulletin, 2.*

Burkill, I. H. (1935). *Dictionary of economic products of Malaya.* Kuala Lumpur: Ministry of Agriculture.

Camara-Leret, R., Paniagua-Zambrana, N., Balslev, H., & Macia, M. J. (2014). Ethnobotanical knowledge is vastly under – documented in northwestern South America. *PLoS One, 9*(1), 14. http://dx.doi.org/10.1371/journal.pone.0085794. e 85794.

Campbell, D., Fordyce, E., Grebneff, E., & Maxwell, P. (1991). Coconuts, coconuts, coconuts. *Geological Society of New Zealand Newsletter, 92,* 37–38.

Cappers, R. T. J. (Ed.). (2006). *Roman footprints at Berimike.* Los Angeles, CA, USA: Cotsen Institute of Archaeology, University of California. 229 pp.

Carson, M. J. (2012). History of archaeological study in the Mariana Islands. *Micronesia, 42,* 312–371.

Casson, L. (1989). *The Periplus Maris Erythraei.* Princeton, NJ, USA: Princeton University Press. 93 pp.

Child, R. (1974). *Coconuts* (2nd ed.). London: Longman. 335 pp.

Chiovenda, E. (1921–1923). La culla del cocos. *Webbia, 5,* 199–294 359–449. (cf. Harries, 0000).

Clark, P. U., Dyke, A. S., Shakun, J. D., Carlson, A. E., Hosteter, S. W., & McCabe, A. M. (2009). The last glacial maximum. *Science, 325,* 710–714.

Clark, P. U., Shakun, J. D., Baker, P. A., Bartlein, P. J., Brewer, S., Brook, Ed, et al. (2012). Global climate evolution during the last deglaciation. *Proceedings of the National Academy of Sciences of the United States of America.* http://dx.doi.org/10.1073/pnas.1116619109.

Clark, R. (1991). Fingota/Fangola: shell fish and fishing in Polynesia. In A. Pairley (Ed.), *Man and a half* (pp. 78–83). Auckland, New Zealand: Polynesian Society.

Clement, C. R., Zizumbo-Villareal, D., Brown, C. H., Ward, R. G., Alves-Pereira, A., & Harries, H. C. (2013). Coconuts in the Americas. *Botanical Review, 79,* 342–370.

COGENT. (2010a). *Conserved germplasm catalogue.* Serdang, Selangor DE, Malaysia: Bioversity International, IPGRI – APO.

COGENT. (2010b). *Catalogue of conserved germplasm by country of origin – India.* Serdang, Selangor DE, Malaysia: Bioversity International, IPGRI – APO, 75–85.

Cook, O. F. (1901). The origin and dispersal of the cocoa palm. *Contributions of the US National Herbarium, 7,* 257–293.

Cook, O. F. (1910). History of the coconut palm in America. *Contributions of the US National Herbarium, 14*, 271–342.

Copeland, E. B. (1931). *The coconut* (3rd ed.). London: Macmillan & Co.

Corner, E. J. H. (1966). *The natural history of palms.* London: Weidenfeld and Nicolson. 393 pp.

Couper, R. A. (1952). The spore and pollen flora of the *Cocos*-bearing beds, Mangonui, New Zealand. *Transactions of the Royal Society of New Zealand, Botany, 79*, 340–348.

Couvreur, T. L. P., Forest, F., & Baker, W. J. (2011). Origin and global diversification patterns of tropical rain forests: inferences from a complete genus-level phylogeny of palms. *BMC Biology, 9*, 44. 12 pp. http://www.biomedcentral.com/1741-7007/9/44.

Crabtree, D. R. (1987). Angiosperms of the northern Rocky Mountains: Albian to companion (Cretaceous) megafossil flora. *Annals of the Missouri Botanical Garden, 74*, 707–747.

Daghlian, C. P. (1981). A review of the fossil record of monocotyledons. *Botanical Review, 47*, 517–555.

Darwin, C. (1889). *The structure and distribution of coral reefs* (3rd ed.). UK: Darwin Online (First edition, 1842).

De Condolle, A. (1886). *Origin of cultivated plants* (2nd ed.). New York: Reprint. Hafner Publishing Company. 468 pp. First edition, 1882.

Dennis, J. V., & Gunn, C. R. (1971). Case against trans-Pacific dispersal of the coconut by ocean currents. *Economic Botany, 25*, 407–413.

Devakumar, K., Jayadev, K., Rajesh, M. K., Chandrasekhar, A., Manimekalai, R., Kumaran, P. M., et al. (2006). Assessment of genetic diversity of Indian coconut accessions and their relationship to other cultivars using microsatellite markers. *Plant Genetic Resources Newsletter, 145*, 38–45.

Doebley, J. F., Gaut, B. S., & Smith, B. D. (2006). The molecular genetics of crop domestication. *Cell, 127*, 1309–1321.

Dowe, J. L., & Smith, L. T. (2002). A brief history of the coconut palm in Australia. *Palms, 46*, 134–138.

Doyle, J. J., & Gaut, B. S. (2000). Evolution of genes and taxa: a primer. *Plant Molecular Biology, 42*, 1–23.

Dransfield, J., Flenley, J. R., King, S. M., Harkness, D. D., & Rapu, S. (1984). A recently extinct palm from the Easter island. *Nature, 312*, 750–752.

Dransfield, J., & Uhl, N. W. (1998). Palmae. In K. Kubitski (Ed.), *The families and genera of vascular plants Flowering plants. Monocotyledons: Vol. IV.* (pp. 306–389). Berlin: Springer.

Dransfield, J., Uhl, N. W., Asmussen, C. B., Baker, W. J., Harley, M. M., & Lewis, C. E. (2008). *Genera palmarum. The evolution and classification of palms (GP II).* Kew, UK: Kew Publishing. 732 pp.

Dwyer, R. E. P. (1938). Coconut improvement by seed selection and plant breeding. *New Guinea Agricultural Gazette, 4*, 24–102.

Dy, R. T., & Reyes, S. (2008). *The Philippine coconut industry: Performance, issues and recommendations.* Manila, The Philippines: USAID. 48 pp.

Eiserhardt, W. L., Swenning, J.-C., Kissling, W. D., & Balslev, H. (2011). Geographical ecology of the palms (Arecaceae): determinants of diversity and distribution across spatial scales. *Annals of Botany, 108*, 1391–1416.

Endt, R., & Hayward, B. W. (1997). Modern relatives of New Zealands' fossil coconuts in high altitude South America. *Geological Society of New Zealand Newsletter, 113*, 67–71.

FAOSTAT. http://faostat.fao.org. Various dates.

FBS. http://fbs.fao.org. Various dates.

Foale, M. (2003). *The coconut odyssey. The bounteous possibilities of the tree of life.* Canberra, Australia: Australian Centre for International Agricultural Research. 134 pp.

Foale, M., & Harries, H. H. (2011). *Coconut (Cocos nucifera).* Speciality Crops for Pacific Island Agroforestry. 24 pp. http://agroforestrynet/seps.

Fosberg, F. R. (1962). A theory on the origin of the coconut. In *The Symposium on the Impact of Man on Humid Tropics Vegetation, Goroka* (pp. 73–75). Canberra, Australia: PNG Government Printer.

Fosberg, F. R., & Corwin, G. (1958). A fossil flora from Pagan, Mariana Islands. *Pacific Science*, *12*, 3–15.

Freitas, C., Meerow, A. W., Pintaud, J.-C., Henderson, A., Noblick, L., Coste, F. R. C., et al. (2016). Phylogenetic analysis of *Attalea* (Arecaceae): insights into historical biogeography of a recently biodiversified Neotropical land group. *Botanical Journal of the Linnaean Society*, *182*, 287–302.

Fremond, Y., Ziller, R., & de Nuce de Lamothe, M. (1966). *Le cocotier*. Paris: Monssonneure & Larose.

Friis, E. M., Penderson, K. R., & Crane, P. R. (2004). Aracaceae from the early Cretaeous of Portugal: evidence on the emergence of monocotyledons. *Proceedings of the National Academy of Sciences of the United States of America*, *101*, 16565–16570.

Futey, M. K., Gandolfo, M. A., Zamaloa, M. C., Cuneo, R., & Cladera, G. (2012). Aracaceae fossil fruits from the Palaeocene of Patagonia, Argentina. *Botanical Review*, *78*, 205–234.

Gandhi, M., & Singh, Y. (1989). *On the mythology of Indian plants*. Calcutta, India: Rupa and Co. 175 pp.

Gangolly, S. R., Satyabalan, K., & Pandalai, K. M. (1957). Varieties of the coconut. *Indian Coconut Journal*, *10*, 3–28.

Gaut, B. S. (2002). Evolutionary dynamics of grass genomes. *New Phytologist*, *154*, 15–28.

Gibb, H. A. R. (1929). *Ibn Batuta. Travels in Asia and Africa, 1325–1354* (Reprint ed.). New Delhi: Manohar Publishers.

Glassman, S. (1987). *Revision of the palm genus Syagrus Mart, and other selected genera of the Cocos Alliance*. Urbana, IL, USA: University of Illinois Press. 230 pp.

Glassman, S. (1999). *A taxonomic treatment of the palm subtribe attaleinae (tribe Cocoseae)*. Urbana, IL, USA: University of Illinois Press. 414 pp.

Glover, I., & Bellwood, P. (2004). *Southeast Asia: From prehistory to history*. Abington, UK: Routledge Curzon.

Golsen, J. (1989). The origins and development of New Guinea agriculture. In D. R. Harries, & G. C. Hillman (Eds.), *Foraging and farming: The evolution of plant exploitation* (pp. 678–687). London: Unwin Hyman.

Gomez-Navarro, G., Jaramilo, C., Herrera, F., Wing, S. L., & Callejas, R. (2009). Palms (Arecaceae) from a Palaeocene rainforest of northern Columbia. *American Journal of Botany*, *96*, 1300–1312.

Gosden, C. (1992). Production systems and the colonization of the western Pacific. *World Archaeology*, *24*, 55–69.

Gosden, C. (1995). Arboriculture and agriculture in coastal Papua New Guinea. *Antiquity*, *65*(265), 807–816.

Green, R. C. (1991). Near and remote Oceania. In A. Pawley (Ed.), *Man and a half* (pp. 491–502). Auckland: The Polynesian Society.

Gruezo, W. Sm. (1990). Fruit component analysis eight "wild" coconut populations in the Philippines. *Philippine Journal of Coconut Studies*, *15*, 6–15.

Greuzo, W. Sm., & Harries, H. C. (1984). Self-sown, wild type coconuts in the Philippines. *Biotropica*, *16*, 140–147.

Gunn, B. F. (2004). Phylogeny of the Cocoeae (Arecaccae) with emphasis on *Cocos nucifera*. *Annals of the Missouri Botanical Garden*, *91*, 505–525.

Gunn, B. F., Baudouin, L., Buele, T., Ilbert, P., Duperray, C., Crisp, M., et al. (2015). Ploidy and domestication are associated with genome size variation in palms. *American Journal of Botany*, *102*, 1625–1633.

Gunn, B. F., Baudouin, L., & Olsen, K. M. (2011). Independent origins of cultivated coconut in the Old World tropics. *PLoS One, 6*(6), ez 1143. http://dx.doi.org/10.1371/journal. Pone. 0021143.

Gupta, S. M. (1996). *Plants in Indian temple art.* Delhi: B.R. Publishing Corporation. 241 pp.

Hahn, W. J. (2002a). A phylogenetic analysis of the Arecoid line of palms based on plastid DNA sequence data. *Molecular Phylogenetics and Evolution, 23*, 189–204.

Hahn, W. J. (2002b). A molecular phylogenetic study of the Palmae (Arecaccae) based on atpB, rbcL, and 18S nrDNA sequences. *Systematic Biology, 51*, 92–112.

Hall, R. (2002). Cenozoic geological and plate tectoric evolution of Southeast Asia and the south-west Pacific. *Journal of Asian Earth Sciences, 20*, 353–431.

Handover, W. P. (1919). The dwarf coconut. *Malayan Agricultural Journal, 8*, 295–297.

Harlan, J. R. (1971). Agricultural origins: centres and noncentres. *Science, 174*, 468–474.

Harlan, J. R. (1992). *Crops and man* (2nd ed.). Madison, WI, USA: American Society of Agronomy.

Harlan, J. R., & De Wet, J. M. J. (1965). Some thoughts about weeds. *Economic Botany, 19*, 16–24.

Harlan, J. R., & De Wet, J. M. J. (1971). Towards a rational classification of weeds. *Taxon, 20*, 509–517.

Harley, M. M. (2006). A summary of the fossil records of the Arecaceae. *Botanical Journal Linnean Society, 151*, 39–67.

Harley, M. M., & Baker, W. J. (2001). Pollen aperture morphology in Arecaceae: application within phylogenetic analysis and a summary of the fossil record of palm-like pollen. *Grana, 40*, 45–77.

Harries, H. C. (1977). The Cape Verde region (1499–1549), the key to coconut culture in the Western Hemisphere? *Turrialba, 27*, 227–231.

Harries, H. C. (1978). Evolution, dissemination, and classification of *Cocos nucifera. Botanical Review, 44*, 265–320.

Harries, H. C. (1981a). Practical identification of coconut varieties. *Oleagineux, 36*, 63–72.

Harries, H. C. (1981b). Germination and taxonomy of the coconut palm. *Annals of Botany, 48*, 873–883.

Harries, H. C. (1982). Coconut varieties. *Indian Coconut Varieties Indian Coconut Journal, 12*(11), 1–8.

Harries, H. C. (1990). Malesian origin for a domestic *Cocos nucifera.* In P. Baas, et al. (Eds.), *The plant diversity of Malesia* (pp. 351–357). Dordrecht, The Netherlands: Springer.

Harries, H. C. (1992). Biogeography of the coconut, *Cocos nucifera* L. *Principes, 36*, 155–162.

Harries, H. C. (1995). Coconut (*Cocos nucifera* L., Palmae). In J. Smartt, & N. W. Simmonds (Eds.), *Evolution of crop plants* (2nd ed.) (pp. 389–394). Harlow, UK: Longman Scientific and Technical, Burnt Mill.

Harries, H. C. (2001). Coconut. In F. Last (Ed.), *Tree crop ecosystems.* Amsterdam: Elsevier.

Harries, H. C. (2012). Key to coconut cultivation on the American Pacific coast: the Manila–Acapulco galleon route (1565–1815). *Palms, 56*, 72–77.

Harries, H. C. (February 2015). *Personal communication* (06).

Harries, H., Baudouin, L., & Cardena, R. (2004). Floating, boating, and introgression: molecular techniques and the ancestry of coconut palm populations. *Ethnobotany Research and Applications, 2*, 37–53.

Harries, H. C., & Clements, C. R. (2014). Long-distance dispersal of the coconut palm by migration within coral atoll ecosystem. *Annals of Botany, 113*, 565–570.

Harris, W. (1995). Early agriculture in New Guinea and the Torres Straits divide. *Antiquity, 69*(265), 1–7.

Harvard Health Letter. (January 01, 2014). *Coconut oil: Supervillain or superfood?.* www.health.harvard.edu/heart.

Hather, J. G. (1992). The archaeobotany of subsistence in the Pacific. *World Archaeology, 24*, 70–81.

Hawley, J. C. (Ed.). (2008). *India in Africa, Africa in India*. Bloomington, IN, USA: Indiana University Press.

Hayward, B. W., Moore, P. R., & Gibson, R. W. (1960). How warm was the late oligocene in New Zealand? Coconuts, reef corals and large foraminifera. *Geological Society of New Zealand Newsletter, 90*, 39–41.

Henderson, A., Galeano, G., & Bernal, R. (1995). *A field guide to the palms of the Americas*. Princeton, NJ, USA: Princeton University Press.

Heyerdahl, J. (1950). *The Kon-Tiki expedition*. England: Penguin Books, Harmondworth. 235 pp.

Higham, C. (2003). Languages & farming dispersals. In P. Bellwood, & C. Renfrew (Eds.), *Examing the farming/language dispersal hypothesis* (pp. 223–232). Cambridge, UK: McDonald Institute of Archaeological Sudies.

Hill, A. W. (1929). The original home and mode of dispersal of the coconut. *Nature, 124*, 133–134, 151–153.

Hill, A. W., & van Leeuwen, W. D. (1933). Germinating coconuts on a new volcanic island, Krakatoa. *Nature, 132*, 674.

Hoaper, R. (1994). Reconstructing proto-polynesian fish names. In A. K. Pawley, & M. D. Ross (Eds.), *Austronesian terminologies. Continuity and change. Pacific linguistics C-127* (pp. 185–230). Canberra, Australia: Australian National University. (cf. Kirch, 2002).

Hossfeld, P. S. (1965). Radiocarbon dating and palaeo-ecology of the Aitape fosset human remains. *Proceedings of the Royal Society of Victoria, 78*, 161–165.

von Humboldt, A. (1853). (cf., Balick, 1984).

IPGRI. (1995). *Descriptors for coconut (Cocos nucifera L.)*. Macarese, Rome, Italy: IPGRI. 61 pp.

IPNI. (2015). *International plant names index*. Kew, UK: Royal Botanic Garden. www.kew.org/wcsp.

Jacob, K. C. (1941). A new variety of coconut palm (*Cocos nucifera* L. var. spicata K.C. Jacob). *Journal of the Bombay Natural History Society, 41*, 906–907.

Jacob, P. M., & Krishnamoorthy, B. (1981). Observations on the coconut genepool in Lakshadweep Islands. In *Placrosym V* (pp. 3–8). Kasaragod, Kerala, India: C.P.C.R.I.

Janssen, T., & Bremer, K. (2004). The age of major monocot groups inferred from 800+ rbcL sequences. *Botanical Journal of the Linnean Society, 146*, 385–390.

Jerard, B. A., Rajesh, M. K., Thomas, R. J., Niral, V., & Samsudeen, K. (2016). *The island ecosystems host rich diversity in coconut: Evidence from Minicoy Islands, India* (Submitted for publication).

John, C. M., & Narayana, G. V. (1942). The male coconut tree. *Madras Agricultural Journal, 30*, 351–352.

John, C. M., & Narayana, G. V. (1949). Varieties and forms of the coconut. *Indian Coconut Journal, 3*, 209–225.

John, C. M., & Satyabalan, K. (1955). A note on the important coconut varieties of the Laccadive Islands. *Indian Coconut Journal, 8*, 65–73.

Johnson, D. V. (2010). *Tropical palms. Non–wood forest products #10/Rev.1*. Rome: FAO. 256 pp.

Jones, A. M. (1959). Indonesia and Africa: the xylophone as a culture indicator. *Journal of the Royal Anthropological Institute, 89*, 155–168.

Jones, T. L., Storey, A. A., Matisoo-Smith, E. A., & Ramirez-Aliaga, J. M. (Eds.). (2011). *Polynesians in America. Pre-columbian contracts with the new world*. New York: Alta Mira Press. 359 pp.

Kato, S., Kosaka, H., & Hara, S. (1928). On the affinity of rice varieties as shown by the fertility of rice plants. *Science Bulletin, Faculty of Agriculture, Kynshu Imperial University, 3*, 132–142.

Kaul, K. N. (1951). A palm fruit from Kapurdi (Jodhpur, Rajasthan Desert), *Cocos sahnii* sp. nov. *Current Science, 20*, 138.

Kinaston, R., Buckley, H., Valentin, F., Bedford, S., Spriggs, M., Hawkins, S., et al. (2014). Lapita diet in remote Oceania: new staple isotope evidence from the 3000-year old Teouma site, Efate island, Vanuau. 23 pp. *PLoS One, 9*(3), e 90376.

King, R. C., Stansfield, W. D., & Mulligan, P. K. (2006). *A dictionary of genetics*. New York, NY, USA: Oxford University Press. 596 pp.

Kirch, P. V. (1997). *The Lapita peoples. Ancestors of the oceanic world*. Oxford, UK: Blackwell Publishers. 353 pp.

Kirch, P.V. (2000). *On the road of the winds*. Berkeley, CA, USA: University of California Press. 424 pp.

Kirch, P.V. (2015). *Personal correspondence*.

Kirch, P.V., & Yen, D. E. (1982). *Tikopia. The prehistory and ecology of a Polynesian outlier*. Bulletin # 238. Honolulu, HI, USA: B.P. Bishop Museum. (cf. Kirch, 2000).

Kissling, W. D., Eiserhardt, W. L., Baker, W. J., Borchsenius, F., Couvreur, T. L. P., Balslev, H., et al. (2012). Cenozoic imprints on the phylogeneitc structure of palm species assemblages worldwide. *Proceedings of the National Academy of Sciences of the United States of America, 109*, 7379–7384.

Kitalong, A. H., Balick, M. J., Rehuler, F., Besebes, M., Hanser, S., Sooladaob, K., et al. (2011). Plants, people, and culture in the villages of Oikull and Ibobang, Republic of Palau. *Terra Australis, 35*, 63–85.

Kraus, B. H. (1974). *Ethnobotany of the Hawaiians*. Lyon Arboretum Lecture #5. Hawaii, USA: University of Hawaii, Kohala, Centre.

Krishnakumar, V., Jerard, B. A., Josephrajkumar, A., & Thomas, G. V. (2014). *Four decades of CPCRI interventions on coconut–based island ecosystem at Minicoy, Lakshadweep*. Kasaragod, Kerala, India: Central Plantation Crops Research Institute. 27 pp. Technical Bulletin #86.

Krishnamoorthy, B., & Jacob, P. M. (1981). *Fruit component analysis in Lakshadweep coconuts. Placrosym V*. Kasaragod, India: Central Plantation Crops Research Institute, 180–183.

Kwak, M., Velasco, D., & Gepts, P. (2008). Mapping homologous sequences for determinacy and photoperiod sensitivity in common bean, *Phaseolus vulgaris*. *Journal of Heredity, 99*, 283–291.

Ladizinsky, G. (1999). *Plant evolution under domestication*. Dordrecht, The Netherlands: Kluwer. 254 pp.

Lakhanpal, R. N. (1970). Tertiary floras of India and their bearing on the historical geology of the region. *Taxon, 19*, 675–694.

Lakhanpal, R. N., & Bose, M. N. (1951). Some tertiary leaves and fruits of the Guttiferae from Rajasthan. *Journal of the Indian Botanical Society, 30*, 132–136. (cf. Lakhanpal, 1970).

Lear, C. H., Bailey, T. R., Pearson, P. N., Coxall, H. K., & Rosenthal, Y. (2008). Cooling and ice growth across he Eocene–Oligocene transition. *Geology, 36*, 251–254.

Lebraun, P., Berger, S. A., Hodgkin, T., & Baudouin, L. (2005). Biochemical and molecular methods for characterizing biodiversity. In P. A. Batugal, V. R. Rao, & J. Oliver (Eds.), *Coconut genetic resources* (pp. 225–251). Serdang, Malaysia: IPGRI – APO.

Lebraun, P., N'cho, Y. P., Seguin, M., Grivet, L., & Baudouin, L. (1998). Genetic diversity in coconut (*Cocos nucifera* L.) revealed by RFLP markers. *Euphytica, 101*, 103–108.

Lepesme, P. (1947). *Les insectes des palmiers*. Paris: Lechevalier. (cf. Child, 1974).

Lepofsky, D., Harries, H. C., & Kellum, M. (1992). Early coconuts in Mo'orea, French Polynesia. *Journal of the Polynesian Society, 101*, 299–308.

Lewis, C. E., & Doyle, J. J. (2001a). Phylogenetic analysis of the tribe Areceae (Arecaceae) using two low-copy number genes. *Plant Systematics and Evolution, 236*, 1–17.

Lewis, C. E., & Doyle, J. J. (2001b). Phylogenetic utility of the nuclear gene malate synthase in the palm family (Arecaceae). *Molecular Phylogenetics and Evolution, 19*, 409–420.

Linnaeus, C. (1753). *Cocos*, 1188. http://bioversitylibrary.org/page 359 209. Downloaded 02 January 2016.

Liyanage, D. V. (1958). Varieties and forms of the coconut palm grown in Ceylon. *Ceylon Coconut Quarterly, 9*, 1–10.

Mabberley, D. J. (2008). *Mabberley's plant book* (3rd ed.). Cambridge, UK: Cambridge University Press. 1021 pp.

Mahabale, T. S. (1958). Resolution of the artificial palm genus, palmoxylon: a new approach. *Palaeobotanist, 7*, 76–83.

Mahabale, T. S. (1978). The origin of the coconut. *Palaeobotanist, 25*, 238–248.

Maloney, B. K. (1993). Palaeo-ecology and the origin of the coconut. *Geojournal, 31*, 355–362.

Marechal, H. (1928). Observations and preliminary experiments on the coconut with a view to developing improved seed-nuts for Fiji. *Agricultural Journal Fiji, 1*, 16–45.

Massal, E., & Barrau, J. (1956). Pacific subsistence crops. *Coconut. SPC Quarterly Bulletin, 6*(2), 10–12.

Mathews, P., & Gosden, C. (1997). Plant remains from water-logged sites in Arawe island, PNG. *Economic Botany, 51*, 121–131.

Matisoo-Smith, E. (2007). Radiocarbon and DNA evidence for a pre-Columbian introduction of Polynesian chickens to chile. *Proceedings of the National Academy of Sciences of the United States of America, 104*, 10335–10339.

Maury, C. J. (1930). *O Cretageo da Parahyba do Norte. As floras do Cretaceo Superior da America do Sul* LMonograph Service Geological Museum Brasil No. 8. 305 pp. (cf. Dransfield et al., 2008).

Mayuranathan, P. V. (1938). The original home of the coconut. *Journal of the Bombay Natural History of Society, 40*(174–182), 776.

McCormack, G. (2005). *The origin of the coconut palm.* http://cookislands.bishopmuseum.org.

McKillop, H. (1996). Prehistoric Maya use of native palms. In S. L. Fedick (Ed.), *The managed mosaic: Ancient Maya agriculture and resource use* (pp. 278–296). Provo, UT, USA: University of Utah Press.

Meerow, A. W., Noblick, L., Borrone, J. W., Couvreur, T. L. P., Mauro-Herrera, M., Kuhn, D. N., et al. (2009). Phylogenetic analysis of seven WRKY genes across the palm subtribe Attaleinae (Aracaceae) identifies Syagrus as sister group of the coconut. *PLoS One, 4*(10), e7353.

Meerow, A. W., Noblick, L., Salas-Leiva, D. E., Sanchez, V., Francisco-Ortega, J., Jestrow, B., et al. (2014). Phylogeny and historic biogeography of cocosoid palms (Aracaceae, Arecoideae, Cocoseae) inferred from sequences of six WRKY gene family loci. *Cladistics*, 1–26. http://dx.doi.org/10.1111/cla.12100.

Menon, K. P. V., & Pandalai, K. M. (1958). *The coconut palm. A monograph.* Ernakulam, India: Indian Central Coconut Committee. 384 pp.

Merrill, E. D. (1954). The botany of cook's voyages. *Chronica Botanica, 14*, 161–384.

Mir, R. R., Kudapa, H., Srikanth, S., Saxena, R. K., Sharma, A., Azani, S., et al. (2014). Candidate gene analysis for determinacy in pigeon pea, *Cajanus cajan. Theoretical and Applied Genetica, 127*, 2663–2678.

Mokhtar, G. (Ed.). (1981). General history of Africa *Vol. II. Ancient Civilizations of Africa* (785 pp.). Paris: UNESCO.

Moore, H. E. (1973). The major groups of palms and their distribution. *Gentes Herbarum, 11*, 27–160.

Moore, H. E., & Uhl, N. W. (1982). Major trends of evolution in palms. *Botanical Review, 48*, 1–69.

Morcotte-Rios, G., & Bernal, R. (2001). Remains of palms (Palmae) at archaeological sites in the New World. A review. *Botanical Review, 67*, 309–350.

Morley, R. J. (2000). *Origin and evolution of tropical rainforests.* Chichester, NY, USA: Wiley. (cf. Dransfield et al., 2008).

Mueller, F. von (1867). Australian vegetation, indigenous or introduced. *Journal of Botany, 5*, 160–174. (cf. Dowe & Smith, 2002).

Murdoch, G. P. (1959). *Africa: Its people and their culture history.* New York, USA: McGraw-Hill, 456.

Narayana, G. V., & John, C. M. (1949). Varieties and forms of the coconut. *Madras Agricultural Journal, 36*, 349–366.

Nayar, N. M. (1973). Origin and cytogeneties of rice. *Advances in Genetics, 17,* 153–292.

Nayar, N. M. (1978a). *Origin of the coconut palm.* Paper presented at the International Symposium on Coconut Research and Development, December 1976. Kasaragod, Kerala, India: Book of Abstracts. Central Plantation Crops Research Institute.

Nayar, N. M. (1978b). Coconut research and development. *Principes, 22,* 70–74.

Nayar, N. M. (2007). R & D imperatives in coconut. *Indian Coconut Journal, 38*(7), 2–6.

Nayar, N. M. (2010). The bananas: botany, origin, dispersal. *Horticultural Reviews, 36,* 117–164.

Nayar, N. M. (2014a). *Origins and phylogeny of rices.* Amsterdam, The Netherlands: Elsevier. 305 pp.

Nayar, N. M. (August 22–23, 2014b). *Substainability and profitability of coconuts, arecanut, and cocoa farming technological advances and the way forward. inaugural lecture.* Kasaragod, India: CPCRI (in press).

Neal, V. E., & Trevick, S. A. (2008). The age and origin of Pacific islands: a geological overview. *Philosophical Transactions of the Royal Society B, 362,* 3293–3308.

Nicols, D., & Johnson, L. (2008). *Plants and the K-T boundary.* Cambridge: Cambridge University Press. (cf. Meerow et al., 2014).

Niral, V. (2012). Personal communication.

Noblick, L. (2013). Syagrus. An overview. *The Palm Journal, 205,* 4–30.

Noblick, L. R., & Meerow, A. W. (2015). The transfer of the genus Lytocaryum to Syagriss. *The Palms, 59,* 57–62.

Ohler, J. G. (1984). *Coconut, tree of life.* Rome: FAO. 446 pp.

Oka, H. I. (1988). *Origin of cultivated rice.* Amsterdam, The Netherlands: Elsevier.

Omar, A. B. H. (1919). Races of coconut palm. *The Garden's Bulletin, 2,* 143.

Pan, A. D., Jacobs, B. F., Dransfield, J., & Baker, H. W. (2006). The fossil history of palms in Africa and new records from late oligocene of north-western Ethiopia. *Botanical Journal Linnean Society, 151,* 69–81.

Parham, J. W. (1960). *Coconut and breadfruit surveys in the South Pacific region. Tech Info. Bull. #1.* Noumea, New Caledonia: South Pacific Commission. (cf. Harries, 1978).

Parkes, A. (1997). Environmental change and the impact of Polynesian colonization. sedimentary records from central Polynesia. In P.V. Kirch, & T. L. Hunt (Eds.), *Historical ecology in the Pacific islands: Prehistoric environmental and landscape change* (pp. 166–199). New Haven, CT, USA: Yale University Press.

Parkes, A., & Flenley, J. R. (1990). The Hull university Moorea expedition, 1985. *Miscellaneous Series # 37.* Hull: University of Hull.

Patel, J. S. (1938). *The coconut. A monograph.* Madras, India: Government Press. 313 pp.

Perera, L., Baudouin, L., & Mackay, I. (2016). SSS markers indicate a common origin of self-pollinating dwarf coconut in south-east Asia under domestication. *Sciential Horticulturae, 221,* 255–262.

Piperno, D. R., & Pearsall, D. M. (1998). *The origins of agriculture in the lowland neotropics.* New York: Academic Press. (cf. Clement et al., 2013).

Possehl, G. L. (1997). *Harappan civilization. A contemporary perspective.* New Delhi: Vistar Publications.

Prasad, M., Khara, E. G., & Singh, S. K. (2013). Plant fossils from the Deccan intertrappean sediments of Chhindwara district, M.P., India. *Journal of the Palaentological Society of India, 58,* 229–250.

Prebble, M., & Dowe, J. L. (2008). The Late Quaternary decline and extinction of palms on Oceanic Pacific islands. *Quaternary Science Reviews, 27,* 2546–2567.

Prigge, V., Lagenberger, G., & Martin, K. (2005). *Ethnobotanical survey among farmers in Leyte, Philippines, and comparison with indigenous Filipino plant lore.* Paper read at Conference on International Agricultural Research for Development. Stuttgart, Germany: University of Hohenheim.

Purseglove, J. W. (1968). The origin and distribution of the coconut. *Tropical Science, 10,* 190–199.

Purseglove, J. W. (1985). *Tropical crops: Monocotyledons* (2nd ed.). London: Longman.

Rajendran, R. (1991). *Artocarpus altilis* (Parkinson) Fosberg. In *Prosea – 2: Edible fruits and nuts* (pp. 83–86). Wageningen, The Netherlands: Prosea Foundation.

Ramanujam, C. G. K. (2004). Palms through the ages in southern India – a reconnaissance. *Palaeobotanist, 53*, 1–4.

Rana, R. S., Kumar, K., Loyal, R. S., Sahni, A., Rose, K. D., Mussell, J., Singh, H., & Kulshreshta, S. K. (2006). Selachians from the Early Eocene Kapurdi Formation, Barmer district, Rajasthan. *Journal of the Geological Society of India, 67*, 509–522.

Randhawa, M. S. (1964). *The cult of trees and tree worship in Buddhist–Hindu sculpture.* New Delhi: All India Fine Arts and Crafts Society.

Rao, K. P., & Ramanujam, C. G. K. (1975). A palynological approach to the study of Quilon beds of Kerala State in South India. *Current Science, 44*, 730–732.

Rathnambal, M. J., Nair, M. K., Muralidharan, K., Kumaran, P. M., Rao, E. V. V. B., & Pillai, R. V. (1995). *Coconut descriptors – I.* Kasaragod, Kerala, India: Central Plantation Crops Research Institute. 197 pp.

Rathnambal, M. J., Niral, V., Krishnan, M., & Ravi Kumar, N. (2000). *Coconut descriptors–II. Soft copy only.* Kasaragod, Kerala, India: Central Plantation Crops Research Institute.

Raveendranath, T. V., & Ninan, C. A. (1974). A study of the somatic chromosome complement of tall and dwarf coconuts. *Journal of Plantation Crops, 1*, 17–22.

van Rheede Tot Drakenstein, H. (1678–1693). Hortus Indicus Malabaricus. In J. V. Someran, & J. V. Dyck (Eds.), *Latin* (12 Vols). Trivandrum: Amsterdam. English Version Published by the University of Kerala. 2011.

Richardson, J. E., Costion, C. M., & Muellner, A. M. (2012). The Malesian floristic interchange plant migration patterns across Wallace's Line. In D. Gower, et al. (Eds.), *Biotic evolution and environmental change in Southeast Asia.* Cambridge, UK: Cambridge University Press (pp. 138–163). (cf. Baker & Couvreur, 2013b).

Ridley, H. N. (1930). *The dispersal of plants throughout the world.* Ashford, UK: Reeve & Co.

Rigby, J. F. (1995). A fossil *Cocos nucifera* L. fruit from the latest Pliocene of Queensland, Australia. In D. D. Pant (Ed.), *Birbal Sahni Centennial volume* (pp. 379–381). Allahabad, India: Allahabad University, Allahabad, India/South Asian Publishers.

Riley, C. L., Kelley, J. C., Pennington, C. W., & Rands, R. L. (Eds.). (1971). *Man across the sea.* Austin, TX, USA: University of Texas Press.

Sahni, B. (1946). A silicified Cocos-like Palmoxylon (Cocos) sundaram, from the Deccan intertrappean beds. In M. O. P. Iyengar (Ed.), *Commemoration volume. Journal of the Indian Botanical Society* (pp. 361–374).

Samsudeen, K., Jacob, P. M., Niral, V., Kumaran, P. M., Salooja, R., & Moosa, H. (2006). Exploration and collection of coconut germplasm in Kadmat and Amini Islands. *Genetic Resources and Crop Evolution, 53*, 1721–1728.

Samsudeen, K., Jacob, P. M., Rajesh, M. K., Jerard, B. A., & Kumaran, P. M. (2006). Origin and evolution of Laccadive Micro Tall, a coconut cultivar from Lakshadweep Islands of India. *Journal of Plantation Crops, 34*, 220–225.

Santos, G. A. (1983). *Coconut germplasm collection in the Philippines.* Laguma: University of the Philippines, Los Banos, College. 16 pp.

Sauer, J. D. (1967). *Plants and man on the Seychelles coast.* Madison, WI, USA: University of Wisconsm Press. 132 pp.

Sauer, J. D. (1971). A re-evaluation of the coconut as an indicator of human dispersal. In C. L. Riley, J. C. Kelley, C. W. Pennington, & R. L. Rands (Eds.), *Man across the Sea. Problems of pre-Columbian contacts* (pp. 309–319). Austin, TX, USA: University of Texas Press. 551 pp.

Savolainen, V., Anstett, M.-C., Lexer, C., Hutton, I., Clarkson, J. J., Norup, M. V., et al. (2006). Sympatric speciation of palms in an oceanic island. *Nature, 441*, 210–213.

Schoff, W. A. (1912). *Translated and annotated by the Periplus of the Erythsean Sea*. London: Longman, Green, and Co. 323 pp.

Sheriff, A. M. H. (1981). The east African coast and its role in maritime trade. In G. Mokhtar (Ed.), *General history of Africa II. Ancient civilizations* (pp. 551–567). Paris: UNESCO.

Shukla, A., Mehrotra, R. C., & Guleria, J. S. (2012). *Cocos sahnii* Kaul: a *Cocos nucifera* L.-like fruit from the early eocene rain forest of Rajasthan, Western India. *Journal of Biosciences*, *37*, 769–776.

Sidebotham, S. E. (2011). *Berenike and the Ancient Martine spice route*. Berkeley, CA, USA: University of California Press.

Simmonds, N. W. (1966). *Bananas*. Harlow, UK: Longman.

Simmonds, N. W. (Ed.). (1974). *Evolution of crop plants*. Harlow, UK: Longman. 339 pp.

Singh, U. (2009). *A history of ancient and early medieval India*. New Delhi: Pearson. 677 pp.

Singh, H. P., & Rao, M. R. (1990). Tertiary palynology of Kerala basin – an overview. *Palaeobotanist*, *38*, 256–262.

Small, J. K. (1929). The coconut palm – *Cocos nucifera*. *Journal of the New York Botanic Garden*, *30*(355), 153–161, 356, 194–203.

Smith, A. G., Smith, D. G., & Funnell, B. M. (1994). *Atlas of mesozoic and cenozoic coastlines*. Cambridge: Cambridge University Press. 99 pp.

Soltis, D. E., & Soltis, P. S. (2000). Contributions of plant molecular systematics to studies of molecular evolution. *Plant Molecular Biology*, *42*, 45–75.

Soltis, D. E., & Soltis, P. S. (2003). The role of phylogenetics in comparative genetics. *Plant Physiology*, *132*, 1790–1800.

Sowunmi, M. A. (1968). Pollen morphology of the Palmae with special reference to trends in aperture development. *Review of Palaeobotany and Palynology*, *7*, 45–53.

Sowunmi, M. A. (1972). Pollen morphology of the palmae and its bearing on taxonomy. *Review of Paleobotany and Palynology*, *13*, 1–80.

Spriggs, M. (1984). Early coconut remains from the South Pacific. *Journal of the Polynesian Society*, *93*, 71–76.

Srivastava, R., & Srivastava, G. (2014). Fossil fruit of Cocos L. (Arecaceae) from Mastrichian – Damian sediments of central India and its phytogeographical significance. *Acta Palaeobotanica*, *54*, 67–75.

Stebbins, G. L. (1950). *Variation and evolution in plants*. New York, USA: Columbia University Press.

Stebbins, G. L. (1981). Co-evolution of grasses and herbivores. *Annales of the Missouri Botanic Gardens*, *68*, 75–86.

Storey, A. A., Quiroz, D., & Matisso-Smith, E. A. (2011). A reappraisal of the evidence for pre-Columbian introduction of chickens to the Americas. In T. L. Jones, A. A. Storey, E. A. Matisoo-Smith, & J. M. Ramerz-Aliaja (Eds.), *Polynesians in America* (pp. 139–170). Lanham, MD, USA: Alta-Mira Press. 359 pp.

Storey, A. A., Ramirez, J. M., Quiroz, D., Burley, D. V., Addison, D. J., Walter, R., et al. (2007). Radiocarbon and DNA evidence for a pre-Columbian introduction of Polynesian chickens to Chile. *Proceedings of the National Academy of Sciences of the United States of America*, *104*, 10335–10339.

Swaminathan, M. S., & Nambiar, M. C. (1961). Cytology and origin of the dwarf palm. *Nature*, *192*, 85–86.

Thaman, R. R. (1992). The ethnobotany of Pacific islands coastal plants. *Atoll Research Bulletin*, *361*, 1–62.

Thanikaimoni, G. (1966). Contribution a letude palynologique des palmiers. *Travanx de la Section Scientifique et Technique*, *5*, 1–92.

Thanikaimoni, G. (1970). *Institute Francois de Pondichery, India Les Palmiers: polynologic et systematique*. *Travaux de la Section Scientifique et Technique 11*, 1–286.

Thapar, R. (2002). *Early India from the origins to AD 1300*. London: Penguin Academics/Allan Lane. 556 pp.

Thozet, A. (1869). The coconut in Australia. *Journal of Botany*, *7*, 213–214. (cf. Rowe & Smith, 2002).

Tian, Z., Wang, X., Lee, R., Li, Y., Specht, J. E., Nelson, R. L., et al. (2010). Artificial selection for determinate growth habit in soybean. *Proceedings of the National Academy of Sciences of the United States of America, 107*, 8563–8568.

Tomlinson, P. B. (2006). The uniquecess of palms. *Botanical Journal Linnean Society, 151*, 5–14.

Tripathi, R. P., Mishra, S. N., & Sharma, B. D. (1999). *Cocos nucifera*-like petrified fruits from the Tertiary of Amarkantak, MP, India. *Palaeobotanist, 48*, 251–255.

Uhl, N. W., & Dransfield, J. (1987). *Genera Palmarum. Classification of palms based on the work of Harold E Moore Jr.*. Lawrence, KA, USA: International Palm Society.

Uhl, N. W., Dransfield, J., Dans, J. I., Luckow, M. A., Hansen, K. S., & Doyle, J. J. (1995). Phylogenetic relationships among palms: cladistic analyses of morphological and chloroplast DNA restriction variation. In P. J. Rudall, P. J. Cribb, D. F. Cutler, & C. J. Humphries (Eds.), *Monocotyledons: Systematics and evolution* (pp. 623–661). Kew, UK: Royal Botanic Garden.

Uichanco, L. B. (1931). On the palms which are called Cocos and their great usefulness. Original in Spanish by F.I. Alzina (1668) (Translator) *Philippine Agriculturist, 20*, 435–446. (cf. Greuzo, 1990).

Van Leeuwen, W. D. (1933). Germinating coconuts on a new volcanic island, Krakatau. *Nature, 132*, 674–675.

Van der Kaars, W. A. (1990). *Late Quaternary vegetation and climate of Australasia as reflected by the palynology of eastern Indonesian deep-sea piston cores* (Ph.D. thesis). University of Amsterdam. (cf. Maloney, 1993).

Vansina, J. (1990). *Paths in the rainforests*. Madison, WI, USA: University of Wisconsin Press. 428 pp.

van der Veen, M. (Ed.). (1999). *The exploitation of plant resources in Ancient Africa*. Dordrecht, The Netherlands: Kluwer Academic. 283 pp.

Vavilov, N. I. (1950). *The origin, variation, immunity and breeding of cultivated plants* (Vol. 13). Waltham, MA, USA: Chronica and Botanica. 366 pp.

Verin, P. (1981). Madagascar. In G. Mokhtar (Ed.), *General history of Africa II. Ancient civilizations* (pp. 693–713). Paris: UNESCO. Vinay Chand 2016. Personal communication.

Ward, J. V., Athens, J. S., & Holton, C. (1998). *Holocene pollen records from the Pago river valley*. (cf. Kirch, 2002).

Ward, R. G., & Brookfield, M. (1972). The dispersal of the coconut: did it float or was it carried to Panama. *Journal of Biogeography, 19*, 467–480.

Warner, B., & Querke, D. (2007). *A review of the future prospects for the world coconut industry and past research in coconut production and product*. Canberra, ACT 2601, Australia: ACIAR.

Watt, G. (1889). Dictionary of economic products of India. In *Cocos Linn., Cocos nucifera L* (Vol. I) (pp. 415–459). Calcutta: Government Printer.

Whitehead, R. A. (1965). Speed of germination: a characteristic of possible taxonomic significance in *Cocos nucifera* L. *Tropical Agriculture (Trin.), 42*, 369–372.

Whitehead, R. A. (1966). *Sample survey and collection of coconut germplasm in the Pacific islands*. London: HMSO.

Whitehead, R. A. (1968). *Collection of coconut germplasm from the Indian/Malaysian region, Peru and Seychelles Islands*. Rome: FAO.

Whitehead, R. A. (1974). Coconut. In N. W. Simonds (Ed.), *Evolution of crop plants* (pp. 221–224). Beernt Mill, UK: Longman Scientific and Technical. 339 pp.

Wiens, H. J. (1962). *Atoll ecology and environment*. New Haven, CT, USA: Yale University Press. 531 pp.

Wikipedia. Various dates. https://en.Wikipedia.org/wiki/Pacific Ocean. Updated 30 April 2016; Indian Ocean 15.01.2015.

Wikstroem, N., Savolainen, V., & Chase, M. W. (2001). Evolution of the angiosperms: calibrating the family tree. *Proceedings of the Royal Society of London, Series B. Philosophical Transactions of the Royal Society of London. Series B, Biological Sciences, 268*, 2211–2220. www.//FAOSTAT. Various dates.

Williams, H. (1967). Bikini nine years later. *Science Journal, 3*(4), 48–53. (cf. Child, 1974).

Wilmhurst, J. M., Hunt, T. L., Lipo, C. P., & Anderson, A. J. (2011). High-precision radiocarbon dating shows recent and rapid initial human colonization of east Polynesia. *Proceedings of the National Academy of Sciences of the United States of America, 108*, 1815–1820.

Wilson, (1961). (cf. Singh, 2011).

Yen, D. E. (1990). Environment, agriculture, and the coconization of the Pacific. In D. E. Yen, & J. M. J. Mummery (Eds.), *Pacific production systems* (pp. 258–277). Canberra, Australia: Department of Prehistory, Australian National University.

Yen, D. E. (1995). The development of Sahul agriculture with Australia as bystander. *Antiquity, 69*(265), 831–846.

Yen, D. E., & Mummery, J. M. J. (Eds.). (1990). *Pacific production systems.* Canberra, Australia: Department of Prehistory, Australian National University. 277 pp. Occasional Paper # 18.

Zeven, A. C. (1967). *The semiwild oilpalm and its industry in Africa.* Agricultural Research Reports # 689. Wageningen, The Netherlands: PUDOC.

Zeven, A. C., & Zhukovsky, P. M. (1975). *Dictionary of cultivated plants and their regions of diversity.* Wageningen, The Netherlands: PUDOC. 219 pp.

Zhang, S., Sandal, N., Polowide, P. L., Stong, J., & Fobert, R. R. (2003). Proliferating floral organs (Pfa), a Lotus japonicas gene required for specifying floral meristem determinacy and organ identity encodes a F − bar protein. *The Plant Journal, 33*, 607–619.

Zizumbo-Villareal, D. (1996). The history of coconut (*Cocos nucifera* L., 1539–1810. *Genetic Resources and Crop Evolution, 43*, 505–515.

Zizumbo-Villareal, D., Fernandez-Barrera, M., Torres-Hernandez, N., & Colunga-Martin, P. (2005). Morphological variation of fruit in Mexican populations of *Cocos nucifera* L. under in situ and ex situ conditions. *Genetic Resources and Crop Evolution, 52*, 421–434.

Zizumbo-Villareal, D., Hernandez-Roque, F., & Harries, H. C. (2005). Coconut varieties in Mexico. *Economic Botany, 47*, 65–78.

Zohary, D., Hopf, M., & Weiss, E. (2012). *Domestication of plants in the new world* (4th ed.). Oxford, UK: Oxford University Press. 233 pp (First edition, 1987).

FURTHER READING

Bentham, G. (1878). *Flora Australiensis* (Vol. 7). London: L. Reeve and Co. (cf. Dowe & Smith, 2002).

Collins, R. O., & Burns, J. M. (2007). *A history of sub-Saharan Africa.* Cambridge, UK: Cambridge University Press. 406 pp.

Durocher Yvon, F. (1953). The coconut industry of Seychelles. *World Crops, 5*, 437.

Harries, H. C. (1971). Coconut varieties in America. *Oleagineux, 26*, 235–242.

Harries, H. C. (2000). *The Cambridge historical, geographical, and cultural encyclopedia of human nutrition.* Cambridge, UK: Cambridge University Press. Coconut. 20 pp.

Jack, H. W., & Sands, W. N. (1922). The dwarf coconuts in Malaya. *Malayan Agricultural Journal, 10*, 4–12.

Larget, B. R., Kotha, S. K., Dewey, C. N., & Ane', C. (2010). BUCKY, gene-tree/species-tree reconciliation with Bayesian concordance. *Bioinformatics, 26*, 2910–2911.

Lepofsky, D. (1995). A radiocarbon chronology for prehistoric agriculture in the Society Islands, French Polynesia. *Radiocarbon, 37*, 917–930.

Noblick, L. R. (2014). Syagrus an overview. *The Palm Journal, 205*, 3–31.

Noblick, L. R., Lorenzi, H., & Souza, V. C. Four new taxa of acaulescent *Syagrus* (Arecaceae) from Brazil. *Phytotaxa, 188*, 1–10.

Rajesh, M. K., Nagarajan, P., Jerard, B. A., Arunachalam, V., & Dhanapalan, R. (2008). Microsatellite variability of coconut accessions from Andaman and Nicobar Islands. *Current Science, 94*, 1627–1630.

Sankaran, M., Damodaran, V., Singh, D. R., Sakar, I. J., & Jerard, B. (2012). Characterisation and diversity assessment in coconut collections of Pacific Ocean islands and Nicobar islands. *African Journal of Biotechnology*, *11*, 16320–16329.

Schrank, K. (1994). Palynology of soma Formation in northern Somalia. *Palaentographica Abteilung B*, *231*, 63–112. (cf., Baker & Couvreur, 2013a).

Werth, E. (1933). Distribution, origin, and cultivation of the coconut palm. *Berichte der Deutschen botanischen Gesselschaft*, *51*, 301–304 (in German, translated by R. Child).

INDEX

'*Note*: Page numbers followed by "f" indicate figures and "t" indicate tables.'